西安交通大学本科"十二五"规划教材

SOIL MECHANICS

土力学

Liao Hong-jian　Su Li-jun　Li Hang-zhou　Xiao Zheng-hua
廖红建　苏立君　李杭州　肖正华 编著

西安交通大学出版社
XI'AN JIAOTONG UNIVERSITY PRESS

内容提要

本书是土木工程专业的必修课程《土力学》的英文教材。本教材既注重我国土力学课程的知识结构体系，又吸收国外经典土力学英文原版教材的精髓，采用我国现行的规范和标准进行编写。全书重视土力学基本理论和知识、技能的阐述，力求把知识的传授与能力的培养结合起来，内容丰富，条理清晰，系统性和逻辑性较强，便于学生系统学习和深入理解。

本书内容共分 8 章，包括土的基本性质和工程分类、土的渗透性和渗流力、地基中的应力分布、土的压缩性和地基沉降计算、土的抗剪强度、地基承载力、边坡稳定性以及土压力和挡土墙。每章后附有习题和参考文献，附录中列出了土力学常用专业名词英汉对照和习题答案，为学生和教师查阅和学习提供了方便。

本书可作为高等院校土木工程专业《土力学》课程的双语教材，其内容与中文教材的知识体系相呼应，也可作为相关专业（水利水电工程、采矿工程、交通运输工程等）的土力学课程的双语教材，以及研究生和工程科技人员的参考书。

图书在版编目（CIP）数据

土力学 = Soil Mechanics：英文/廖红建等编著. —西安：西安交通大学出版社，2015.7（2016.12 重印）

西安交通大学"十二五"规划教材

ISBN 978-7-5605-7754-8

Ⅰ.①土… Ⅱ.①廖… Ⅲ.①土力学-高等学校-教材-英文 Ⅳ.①TU43

中国版本图书馆 CIP 数据核字（2015）第 187420 号

书　名	土力学
编　著	廖红建　苏立君　李杭州　肖正华
责任编辑	王晓芬
出版发行	西安交通大学出版社
	（西安市兴庆南路 10 号　邮政编码 710049）
网　址	http://www.xjtupress.com
电　话	（029）82668357　82667874（发行中心）
	（029）82668315（总编办）
传　真	（029）82668280
印　刷	陕西奇彩印务有限责任公司
开　本	727mm×960mm　1/16　印张 15.5　字数 274 千字
版次印次	2015 年 12 月第 1 版　2016 年 12 月第 2 次印刷
书　号	ISBN 978-7-5605-7754-8/TU·164
定　价	38.00 元

读者购书、书店添货、如发现印装质量问题，请与本社发行中心联系、调换。

订购热线：（029）82665248　　（029）82665249

投稿热线：（029）82664953

读者信箱：475478288@qq.com

版权所有　侵权必究

Brief Introduction

This book was compiled as a textbook of soil mechanics for undergraduate students in civil engineering. It substantially considered the structure and contents of soil mechanics in China. Moreover, good materials from foreign textbooks of soil mechanics in English were also included. All of the codes and standards in China included in this book are the newest. The authors paid much attention to explain the basic theory, knowledge and skills in soil mechanics, tying to combine delivery of knowledge and improve the ability of students.

There are eight chapters in this book, including: basic characteristics and engineering classification of soils, permeability of soil and seepage force, stress distribution in soil, compression and consolidation of soil, shear strength, bearing capacity, stability of slopes and lateral earth pressure and retaining walls. Exercises and references are provided at the end of each chapter, and answers are given in the appendix for the reference of lecturers. The English-Chinese translation of frequently-used words and expressions in soil mechanics were also provided in the appendix, which may facilitate the students to study this book.

This book can be used as a bilingual teaching textbook of soil mechanics for undergraduate students in civil engineering because its contents are consistent with the current textbooks in Chinese. It can also be used as a bilingual teaching textbook of soil mechanics for undergraduate students in hydraulic, mining and transportation engineering. Master students and professionals in related professions can also use it as a reference book.

Preface

1. Importance of this course

Soil mechanics is a basic compulsory course for undergraduate students in civil engineering. Through this course, students should master the basic theory and skills of soil mechanics related to civil engineering. Soil mechanics is closely related to the construction of civil engineering. Many engineering constructions, including dams, embankments, tunnels, canals and waterways, foundations for bridges, roads, buildings, and solid waste disposal systems, etc., depend on the geological conditions of the site and mechanical behavior of the soil.

The geological conditions in China are complicated. There are various types of soil, the properties of which also vary significantly. Further more, some special soils or regional soils (such as soft soil, collapsible loess, expansive soil, red clay and permafrost, etc.) have special behaviors different from general soils. Therefore, it is necessary to study the engineering characteristics of soils in order to take appropriate engineering measures. There are many geotechnical problems caused by natural factors or human activities, which are involved various engineering activities, such as civil engineering, mining engineering, underground engineering, and so on. The design and construction of these engineering constructions are closely related to the geological environment and mechanical behavior of soils. Therefore, it is necessary to understand various characteristics (such as physical, chemical and mechanical etc.) of soils.

2. Characteristics and engineering background of the course

As an important basic application —oriented course as well as a practical science of engineering, soil mechanics is to study the characteristics of ground and engineering behavior of soils. It is used to analyze and solve the engineering problems encountered in the design and construction of the foundation and problems associated with geomaterials. This course is an important part of the civil engineering disciplines.

Soil mechanics is a practical engineering discipline that takes soils as the research objects, which is a branch of engineering mechanics. Weathered rocks might disintegrate, metamorphose, be carried to a new environment by various natural forces and accumulated or deposited there, which formed soils. Based on the knowledge of mechanics and engineering geology, soil mechanics is used to study the stress, strain, strength and stability of soils related to engineering constructions under the action of external factors (such as load, water, temperature etc.), using the principles of mechanics and geotechnical testing techniques. Therefore, soil mechanics is a very practical engineering science.

The utilization of soils can be traced back to ancient time. Our ancestors used soils as construction material to build burial sites, flood protection, and shelters. Neolithic sites found in Banpo village, Xi'an, China indicated that, at that time, people had been able to use soil pedestal and stone foundation for the simple houses with the consideration of foundation stability problems. Later, in the Qin Dynasty the method of compaction was used to build roads, and wood pile in the Sui Dynasty and lime soil foundation in the Tang Dynasty were used to build towers. During the European industrial revolution, with the construction of large buildings, railway and highway and the development of science, the sporadic theory of soil mechanics was established. The first scientific study of soil mechanics was undertaken by French physicist Charles —Augustin de Coulomb, who published an equation for calculating the shear strength of sands and a theory of earth pressure in 1773. Coulomb's work and another theory of earth pressure published by Scottish engineer William Rankine in 1857 are currently still in use to quantify earth pressures. In 1869, Карлович published the world's first book of foundation engineering. According to the theory of elasticity, J. Boussinesq obtained the analytic solution of three−dimensional stress distribution in a foundation under the action of a concentrated load in 1885. In 1900, C. O. Mohr proposed the soil strength theory. In early twentieth Century, people had accumulated a lot of experience and data in engineering practice, and theoretically discussed the strength, deformation and permeability properties of soil. Thus, soil mechanics gradually formed an independent discipline. In the 1920s, L. Prandtl published the bearing capacity theory. In this period, there was a great development in the theory of slope stability. W. Fellenius improved the analysis method for circular sliding of slopes, known as the Fellenius method of slices or the Swedish method of slices. Based on the practice and theoretical investigation by professionals and researchers in civil engineering for more than a century, in 1925, K. Terzaghi summarized and published the first book of soil mechanics in United States. In 1929, he published the book "Engineering geology" together with other writers. Since then, soil mechanics, engineering geology and foundation engineering had gained continuous development each as an independent science, respectively. Since the first session held in 1936 in USA, the International Conference on Soil Mechanics and Geotechnical Engineering had been held for 18 times till 2013. Researchers from around the world exchanged research experience on this subject during the conference. With the development of practice and the advance in science, more and more theories and techniques were used for the research in soil mechanics. Application and improvement of basic characteristics of soil, effective stress principle, consolidation theory, deformation theory, stability soil mass, soil dynamics, soil rheology, etc. in soil mechanics were the main issues in this phase of the study. In 1954, В. В. Соколовский published a book of "Loose media statics". A. W. Skempton, A. W. Bishop and N. Janbu had made contributions to the effective stress principle and the theory of slope stability. Chinese scholars Wenxi Huang, Zongji Chen, Jiahuan Qian and Zhujiang Shen had made contributions to constitutive relations of soil, clay microstructure and soil rheology, geotechnical earthquake engineering and soil rheology, and constitutive relations of soft soil, respectively.

The composition and engineering geological conditions of the supporting soil for foundations are complex and different from each other. Requirements of engineering geology for the supporting soil are different for different engineering constructions. Therefore, the engineering problems of soils are various. The composition, thickness, physical and mechanical properties, bearing capacity of soils, etc. are the basic conditions for assessing the stability of the soil supporting the foundation. Therefore, failure of the supporting soil is frequently encountered in engineering construction. A famous case history was the instability of the supporting soil for a grain elevator in Transcona, Canada. Another one was the non-uniform settlement of the foundation for the leaning tower of Pisa in Italy.

The Transcona grain elevator completed in September 1913 was 59.44m long, 31m high and 23.47m wide. Within 24 hours after loading the grain elevator at a rate of about 1m of grain height per day, the bin house began to tilt and settle. Fortunately, the structural damage was minimal and the bin house was later restored. No borings were done to identify the soils and to obtain information on their strength.

The tower of Pisa is located in the city of Pisa, Italy. The city is located on the Arno River, northwest to Rome. The Tower is 54m in height and 142000kN in weight. The Pisa Tower was built in several stages from 1173 to 1370. During this period, the construction stopped twice due to the tower inclination. Prior to restoration work performed between 1990 and 2001, the tower leaned at an angle of 5.5 degrees. The tower now leans at about 3.99 degrees. This means that the top of the tower is displaced horizontally 3.9m from where it would be if the structure were perfectly vertical. The tower's tilt started during the construction because that the south side of the soil mass supporting the foundation was too soft to properly support the structure's weight. The plastic deformation of the foundation, creep, falling water tables, etc. accelerated the Tower inclination. Circular excavations were used for unloading at the opposite side and grouting was carried out to reinforce the soil surrounding the foundation. The body of the tower also reinforced to prevent it from collapse.

To guarantee the stability serviceability of a building, the bearing capacity of the supporting soil must meet two basic conditions: strength and deformation. The soil should have sufficient strength to ensure the stability of the ground under loading. On the other hand, the deformation of the ground should not exceed the allowable value of required by the building. Therefore, a good ground for construction generally has higher strength and lower compressibility.

3. Main contents and requirements

The main contents and requirements of each part are as follows:

1) The basic properties and the permeability of soil

Students are required to understand the concept of a three-phase composition of the soil, parameters for physical properties of soil and their relationship, soil permeability, the theory background of Darcy's law and methods for measuring the coefficient of permeability, concepts and calculation of hydrodynamic pressure and critical hydraulic gradient, the main types of seepage failure and its prevention measures, and the engineering classification of soil.

2) Stress distribution in soil

The stress in soils will be re-distributed during the construction, which causes the deformation of the ground. If the induced stress is too large and exceeds the ultimate bearing capacity of soils, it will cause failure of the ground. Therefore, to understand the calculation of stress and deformation is the premise for ensuring the serviceability and safety of buildings. The students are required to master basic concepts of effective stress, pore water pressure, gravity stress and additional stress, the theory of effective stress under both hydrostatic and seepage conditions, the calculation methods for gravity stress, effective stress, foundation pressure and additional stresses.

3) Compression and consolidation of soil

The stress and deformation occur under the loads of buildings, which eventually induced subsidence and non-uniform settlement of the foundation. If the settlement exceeds a certain limit, it will cause deformation, cracking, tilting or even dumping of buildings. Therefore, the calculation of the settlement is an important issue related to safety and stability of a building. Thus students are required to understand the compression characteristics of soils and the consolidation state, the compressive index of soil and its determination method, the layer-wise summation method for calculating the foundation settlement, and the foundation settlement calculation methods for normally consolidated and overconsolidated soils, including calculation methods of one-way seepage consolidation settlement of the foundation. These are the fundamental knowledge for settlement calculation and the foundation design of actual engineering constructions.

4) Shear strength of soil and bearing capacity

The shear strength is one of the important mechanical properties of the soil, which is closely related to the stability and serviceability of buildings. Determination of bearing capacity is the basic content for the design of a foundation. The students are required to understand the Mohr-Coulomb theory and the limit equilibrium condition of soil, the shear strength indexes and test methods for determining these indexes, shear strength properties of soil, failure characteristics of soils, and the calculation methods for critical edge loads and ultimate loads.

5) Slope stability and earth pressure

Earth pressure and slope stability are the problems frequently encountered in engineering construction, which also must be analyzed for the engineering design and construction. The students are required to understand the basic concepts and calculation methods for earth pressure, Rankine's earth pressure theory and Coulomb's earth pressure theory, the types of retaining structures; factors affecting the slope stability, the slope stability analysis for non-cohesive and cohesive soil, the Swedish method of slices and Bishop's method of circular sliding surface.

Contents
SOIL MECHANICS

Preface

Chapter 1

Basic characteristics and engineering classification of soils   1

    1.1 Introduction   2
    1.2 Soil composition and phase relationships   3
    1.3 Soil fabric   11
    1.4 Particle size analysis   13
    1.5 Density properties for granular soils   15
    1.6 Plasticity properties of soils   19
    1.7 Soil compaction   24
    1.8 Engineering classification of soils   26
    Exercises   28
    References   29

Chapter 2

Permeability of soils and seepage force   31

    2.1 Introduction   32
    2.2 Capillary phenomena   33
    2.3 Darcy's law   35
    2.4 Determination of permeability coefficient   39
    2.5 Flow nets   46
    2.6 Seepage force and critical hydraulic gradient   48
    Exercises   52
    References   54

Chapter 3

Stress distribution in soils — 55

- 3.1 Introduction — 56
- 3.2 Stresses due to self weight — 59
- 3.3 Effective stress principle — 62
- 3.4 Contact pressure between the foundation and the ground — 66
- 3.5 Additional stress in ground base — 73
- 3.6 Additional stress in plane problem — 87
- Exercises — 97
- References — 98

Chapter 4

Compression and consolidation of soils — 101

- 4.1 Introduction — 102
- 4.2 Compressibility characteristics — 103
- 4.3 Calculation of settlement of foundation — 109
- 4.4 One dimensional consolidation theory — 112
- Exercises — 118
- References — 120

Chapter 5

Shear strength — 123

- 5.1 Shear resistance — 124
- 5.2 Mohr-Coulomb failure criterion — 126
- 5.3 Shear strength tests — 130
- 5.4 Effective stress paths — 130
- 5.5 Characteristic of shear strength of cohesiveless soils — 132
- 5.6 Characteristics of shear strength of cohesive soils — 136
- Exercises — 141
- References — 141

Chapter 6

Bearing capacity — 143

- 6.1 Introduction — 144

Contents

6.2	Critical edge pressure	146
6.3	Prandtl's bearing capacity theory	149
6.4	Modification of Prandtl's bearing capacity theory	151
6.5	Terzaghi's bearing capacity theory	152
	Exercises	158
	References	158

Chapter 7
Stability of slopes — 161

7.1	Introduction	162
7.2	Analysis of a plane translational slip	164
7.3	Analysis for the case of $\varphi_u = 0$	166
7.4	The method of slices	170
7.5	General methods of analysis	177
	Exercises	180
	References	182

Chapter 8
Lateral earth pressure and retaining walls — 183

8.1	Introduction	184
8.2	Earth pressure of a retaining wall	185
8.3	Earth pressure at rest	188
8.4	Rankine's theory of earth pressure	190
8.5	Coulomb's theory of earth pressure	202
8.6	Application of earth pressure theory to retaining walls	207
8.7	Design of earth-retaining structures	210
	Exercises	216
	References	217

APPENDIX I   English-Chinese translation of frequently-used words   218

APPENDIX II   Answers to the exercises   229

SOIL MECHANICS CHAPTER 1

Basic characteristics and engineering classification of soils

1.1 Introduction

Soil mechanics may be defined as the study of the engineering behaviors of soils, with reference to the design of civil engineering structures made from or in the earth. Nearly all of the civil engineering structures, such as buildings, bridges, highways, tunnels, earth retaining walls, embankments, basements, and subsurface waste repositories, towers, canals, and dams, are constructed in or on the surface of the earth. To perform satisfactorily, each structure must have a proper foundation.

The four basic types of geotechnical structure are illustrated in Fig. 1.1.1, and most of other cases are variations or combinations of them. Foundations (Fig. 1.1.1 (a)) transmit loads to the ground and the basic criterion for design is that the settlements should be relatively small. The variables in design of a foundation are the load F, the size of the base B and the depth D. Foundations may support loads that are relatively small, such as car wheels, or relatively large, such as a power station. Slopes (Fig. 1.1.1 (b)) may be formed naturally by erosion or built by excavation or filling. The basic variables are the slope angle i and the depth H, and the design requirement is that the slope should not be failed by landsliding.

Slopes that are too deep and too steep to stand unsupported can be supported by a retaining wall (Fig. 1.1.1 (c)). The basic variables are the height of the wall H and its depth of burial D, together with the strength and stiffness of the wall and the forces in any anchors or props. The basic requirements for design are complex and involve overall stability, restriction of ground movements and the bending and shearing resistance of the wall. In any structure where there are different levels of water, such as in a dam (Fig. 1.1.1 (d)) or around a pumped well, there will be seepage of water. The seepage causes leakage through a dam and governs the yield of a well and it also governs the variation of pressure in the groundwater.

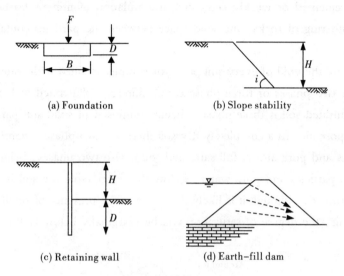

(a) Foundation (b) Slope stability

(c) Retaining wall (d) Earth-fill dam

Fig. 1.1.1 Geotechnical structures

The structures in Fig. 1.1.1 clearly should not fail. There are, however, situation where the material may fail; these include ploughing and flow of mineral ore or grain from storage silo. Solution to problems of this kind can be found using the theories of soil mechanics. Other problems in geotechnical engineering include movement of contaminations from waste repositories and techniques for ground improvement.

1.2 Soil composition and phase relationships

1.2.1 Soil composition

Soil is a particulate material, which means that a soil mass consists of an accumulation of individual particles that are bonded together by mechanical or attractive means, though not as strongly as for rock. To the civil engineering, soil is any un-cemented or weakly cemented accumulation of mineral particles formed by the weathering of rocks, the void space between the particles containing water and/or air.

In nature the void of every soil partly or completely filled with water. Soils can be of either two phases or three phase composition, as illustrated in Fig. 1.2.1. A partially saturated soil is three phases, being composed of solid soil particles, pore water and pore air. In a completely dry soil there are two phases, namely the solid soil particles and pore air. A full saturated soil is also two phases, being composed of solid soil particles and pore water. Below the water table the soil is assumed to be fully saturated, although it is likely that, due to the presence of small volumes of entrapped air, the degree of saturation will be marginally below 100%.

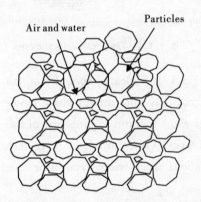

Fig. 1.2.1 Three-phase composition

The water filling the void space may be in a state of rest or in a state of flow. Below the water table the pore water may be static, the hydrostatic pressure depending on the depth below the water table, or may be seeping through the soil under hydraulic gradient, as illustrated in Fig. 1.2.2. If the water is in a state of rest, the methods for solving stability and deformation problems are essentially identical with those for solving similar problems in the mechanics of solids in general. On the other hand, if the water percolates through the voids of the soil, the problems cannot be solved without previously determining the state of stress in the water contained in the voids of the soil. In this case we are obliged to combine the mechanics of solids with applied hydraulics.

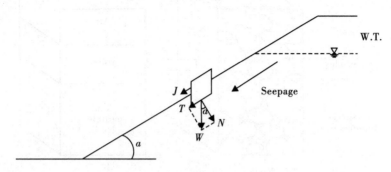

Fig. 1.2.2 Seepage of slope

Significant engineering properties of a soil deposit, such as strength and compressibility, are directly related to or at least affected by basic factors such as how much volume or weight of a bulk soil is solid particles or water or air. Information such as soil density (weight per unit volume), water content, void ratio, degree of saturation is used in calculations to determine the stability of earth slopes, to estimate foundation settlement, and to determine the bearing capacity of foundations. In other words, such information helps to define the condition of a soil deposit for its suitability as a foundation or construction material. For this reason, an understanding of the terminology and definitions relating to soil composition is fundamental to the study of soil mechanics.

◀ 1.2.2 Phase relationships

The components of a soil can be represented by a phase diagram as shown in Fig. 1.2.3. Bulk soil as it exists in nature is a more or less random accumulation of soil particles, water, and air space, as shown in Fig. 1.2.3 (a). For purposes of study and analysis it is convenient to represent this soil mass by a phase or block diagram, with part of the diagram representing the solid particles, part representing water or other liquid, and another part air or other gas, as shown in Fig. 1.2.3 (b).

On the phase diagram, the interrelationships of mass to the volume that make up the soil system being analyzed can be shown. The relationships are summarized in Fig. 1.2.4. The total mass m of the soil volume is taken as the sum of the mass

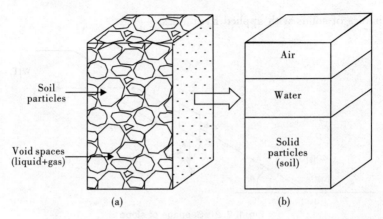

Fig. 1.2.3 Phase diagrams

of solids m_s plus water m_w. The mass of air (in the voids) measured in air (the earth's atmosphere) is zero. The air or other gas may be have a measurable weight, but it would normally be very small compared to the total mass of soil plus water and therefore can be neglected without causing serious error.

The total volume V of the soil bulk includes the volume occupied by solids V_s plus water V_w (or liquid) plus air V_a (or other gas). The total space occupied by water and air may collectively be indicated as the volume of voids.

$$m = m_s + m_w$$
$$V = V_s + V_w + V_a \qquad (1.2.1)$$

The following relationships are defined with reference to Fig. 1.2.4.

(1) The water content (w), or moisture content, is the ratio of the mass of water to the mass of solids in the soil. It is expressed as a percentage, i. e.

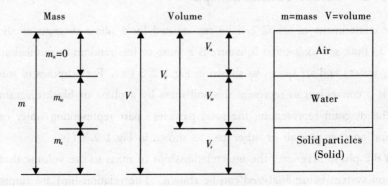

Fig. 1.2.4 Phase relationships

Chapter 1 Basic characteristics and engineering classification of soils

$$w = \frac{m_w}{m_s} \times 100\% \qquad (1.2.2)$$

The water content must be determined by test (drying method). It is determined by weighing a sample of the soil and then drying the sample in an oven at a temperature of 105 −110°C and reweighing.

(2) The bulk density (ρ) of a soil is the ratio of the total mass to the total volume, i. e.

$$\rho = \frac{m}{V} \qquad (1.2.3)$$

Convenient units for density are kg/m³ or Mg/m³. The density of water ρ_w = 1000 kg/m³ or 1.00 Mg/m³.

For a completely dry soil ($S_r = 0$)

$$\rho_d = \frac{m_s}{V} \qquad (1.2.4)$$

For a fully saturated soil ($S_r = 1$)

$$\rho_{sat} = \frac{(m_s + V_v \rho_w)}{V} \qquad (1.2.5)$$

The buoyant density (ρ') is given by

$$\rho' = \frac{(m_s - V_s \rho_w)}{V} \qquad (1.2.6)$$

The unit weight (γ) of a soil is the ratio of the total weight (a force) to the total volume, i. e.

$$\gamma = \frac{W}{V} = \frac{mg}{V} = \rho g \qquad (1.2.7)$$

Convenient units are kN/m³.

(3) The specific gravity of the soil particles (d_s) is given by

$$d_s = \frac{m_s}{V_s \rho_w} = \frac{\rho_s}{\rho_w} \qquad (1.2.8)$$

where ρ_s is the particle density. If the units of ρ_s are Mg/m³ then ρ_s and d_s are numerically equal.

(4) The void ratio (e) is the ratio of the volume of voids to the volume of solids, i. e.

$$e = \frac{V_v}{V_s} \qquad (1.2.9)$$

The porosity (n) is the ratio of the volume of voids to the total volume of the soil, i. e.

$$n = \frac{V_v}{V} \qquad (1.2.10)$$

The e and the n are inter-related as follows:

$$e = \frac{n}{1-n} \qquad (1.2.11)$$

$$n = \frac{e}{1+e} \times 100\% \qquad (1.2.12)$$

(5) The degree of saturation (S_r) is the ratio of the volume of water to the total volume of void space. It is also expressed as a percentage, i. e.

$$S_r = \frac{V_w}{V_v} \times 100\% \qquad (1.2.13)$$

The degree of saturation can range between the limits of zero for a completely dry soil and 1 (or 100%) for a fully saturated soil.

From the definition of void ratio, if the volume of solids is 1 unit then the volume of voids is e units. The mass of solids is then $d_s \rho_w$ and, from the definition of water content, the mass of water is $d_s w \rho_w$. The volume of water is thus $d_s w$. These volumes and masses are represented in Fig. 1.2.5. The following relationships can be obtained.

$$e = \frac{d_s(1+w)\rho_w}{\rho} - 1 \qquad (1.2.14)$$

$$S_r = \frac{d_s w}{e} \qquad (1.2.15)$$

$$\rho = \frac{d_s(1+w)\rho_w}{1+e} \qquad (1.2.16)$$

$$\rho_{sat} = \frac{d_s + e}{1+e} \rho_w \qquad (1.2.17)$$

$$\rho_d = \frac{d_s}{1+e} \rho_w \qquad (1.2.18)$$

Chapter 1 Basic characteristics and engineering classification of soils

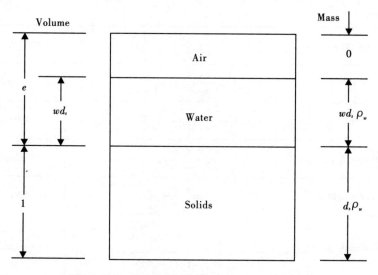

Fig. 1.2.5 Relationship of volumes and masses

Table 1.2.1 shows the formulas of physical indexes.

Table 1.2.1 Practical calculation formulas of physical indexes

Index and the symbol	Expression or conversion formula	Unit
Specific gravity d_s	$d_s = \dfrac{m_s}{m_w} = \dfrac{V_s \rho_s}{V_s \rho_w} = \dfrac{\rho_s}{\rho_w}$	Dimensionless
Water content w	$w = \dfrac{m_w}{m_s} \times 100\%$	%
Density ρ	$\rho = \dfrac{d_s(1+w)\rho_w}{1+e} = \dfrac{d_s + S_r e}{1+e}\rho_w$	g/cm³
Unit weight γ	$\gamma = \dfrac{d_s(1+w)\gamma_w}{1+e} = \dfrac{d_s + S_r e}{1+e}\gamma_w$	kN/m³
Void ratio e	$e = \dfrac{d_s(1+w)\rho_w}{\rho} - 1$, $e = \dfrac{d_s(1+w)\gamma_w}{\gamma} - 1$ $e = \dfrac{d_s \rho_w}{\rho_d} - 1$, $e = \dfrac{d_s \gamma_w}{\gamma_d} - 1$ $e = d_s w$ (当 $S_r = 100\%$ 时)	Dimensionless
Porosity n	$n = \dfrac{e}{1+e} \times 100\%$	%

to be continued

Degree of saturation S_r	$S_r = \dfrac{d_s w}{e}$	%
Dry density ρ_d	$\rho_d = \dfrac{d_s}{1+e}\rho_w$, $\rho_d = \dfrac{\rho}{1+w}$	g/cm³
Dry unit weight γ_d	$\gamma_d = \dfrac{d_s}{1+e}\gamma_w$, $\gamma_d = \dfrac{\gamma}{1+w}$	kN/m³
Saturated density ρ_{sat}	$\rho_{sat} = \dfrac{d_s + e}{1+e}\rho_w$	g/cm³
Saturated unit weight γ_{sat}	$\gamma_{sat} = \dfrac{d_s + e}{1+e}\gamma_w$	kN/m³
Buoyant density ρ'	$\rho' = \dfrac{d_s - 1}{1+e}\rho_w$, $\rho' = \rho_{sat} - \rho_w$ $\rho' = (d_s - 1)(1-n)\rho_w$	g/cm³
Buoyant unit weight γ'	$\gamma' = \dfrac{d_s - 1}{1+e}\gamma_w$, $\gamma' = \gamma_{sat} - \gamma_w$ $\gamma' = (d_s - 1)(1-n)\gamma_w$	kN/m³

Example 1.1

In its natural condition a soil sample has a mass of 2290g and a volume of 1.15×10^{-3} m³. After being completely dried in an oven, the mass of the sample is 2035g. The value of d_s for the soil is 2.68. Determine the water content, bulk density, unit weight, void ratio, porosity and degree of saturation.

Solution

According to Eq. (1.2.2), Eq. (1.2.3) and Eq. (1.2.7), the water content, bulk density and unit weight can be calculated directly by definition.

Water content, $w = \dfrac{m_w}{m_s} = \dfrac{2290 - 2035}{2035} \times 100\% = 12.5\%$

Bulk density, $\rho = \dfrac{m}{V} = \dfrac{2.29}{1.15 \times 10^{-3}} = 1990 \text{kg/m}^3$

Unit weight, $\gamma = \dfrac{mg}{V} = 1990 \times 9.8 = 19.5 \text{kN/m}^3$

The void ratio, porosity and degree of saturation need to use phase relationships, by Eq. (1.2.14), Eq. (1.2.12) and Eq. (1.2.15),

$$\text{Void ratio, } e = \frac{d_s(1+w)\rho_w}{\rho} - 1 = \frac{2.68 \times 1.125 \times 1000}{1990} - 1 = 0.52$$

$$\text{Porosity, } n = \frac{e}{1+e} = \frac{0.52}{1.52} \times 100\% = 34\%$$

$$\text{Degree of saturation, } S_r = \frac{d_s w}{e} = \frac{2.68 \times 12.5\%}{0.52} = 64.5\%$$

Example 1.2

A dry soil is measured with bulk density $\rho = 1.69 \times 10^3 \text{kg/m}^3$, specific gravity $d_s = 2.70$. After a rain, the total volume of the soil doesn't change and the degree of saturation S_r becomes 40%. Calculate bulk density ρ and water content w of the soil after the rain.

 olution

For a dry soil, the dry density ρ_d is numerically equal to the bulk density ρ, that is $\rho_d = 1.69 \times 10^3 \text{kg/m}^3$, then from Eq. (1.2.18),

$$\text{Void ratio } e = \frac{d_s \rho_w}{\rho_d} - 1 = \frac{2.70 \times 1000}{1690} - 1 = 0.6$$

Therefore

$$\text{Water content after the rain } w = \frac{S_r e}{d_s} = \frac{40\% \times 0.6}{2.7} = 8.89\%$$

Bulk density after the rain

$$\rho = \frac{d_s(1+w)\rho_w}{1+e} = \frac{2.7 \times (1+8.89\%) \times 1000}{1+0.6} = 1.84 \times 10^3 \text{kg/m}^3$$

1.3 Soil fabric

In the micro scale, soil fabric is formed by the certain structure of soil particles. When it comes to soil fabric, it refers to particle size, shape, arrangement and especially the mutual connecting condition, which is appearance of soil grain characteristics. Figure 1.3.1 shows the three kinds of basic types of soil fabric.

Crushed stone, gravel, sand and some other coarse grain soil all belong to single grain fabric, which is formed in sedimentation. According to its arrangement situation, it can be divided into two types—the tight and the loose, shown in Fig. 1.3.1 (a). The greater the range of particle sizes present, the tighter the structure can be. The grain of sand or gravel is big and relative surface area and surface energy are quite small compared to its gravity. There is only strong bound water on particle surface and almost no interparticle force among particles. Thus, this kind of soil is also called cohesionless soils.

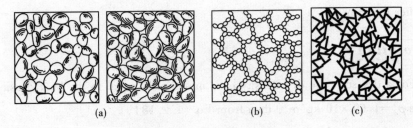

Fig. 1.3.1 Soil Fabric

Silt belongs to honeycomb fabric, shown in Fig. 1.3.1 (b). In the sedimentation process, the fine grained soils once contact the soils which have already sunk, the soil grains will stay at the points of contact and not sink, for which the interparticle force is bigger than the gravity. Thus the honeycomb fabric which has big voids come into being, seen in Fig. 1.3.1(b). This kind of soil is quite unsteady and will produce great deformation under water immersion or external force.

Clay particles belong to flocculent fabric, shown in Fig. 1.3.1(c). Clay ($d <$ 0.005mm), colloidal particles and clay mineral particles ($d <$0.002mm) are very fine, presenting thin-bedded state, which makes this kind of soil with large relative surface area and always suspended in the water. If the interparticle reaction appears attractive force, they easily combine each other and gradually form a little link aggregation of soil grains and make the mass bigger and then sink. When a link touches another, they attract each other, and little by little a big link can be formed. The soil grain of this kind fabric arranges randomly and has big pore space, so its strength is low, compressibility is high and it is sensitive to disturbance. But the linkage strength among soil grains is becoming bigger and bigger for compaction and cementation, which is the main source of cohesive force in clay.

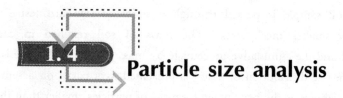

Particle size analysis

The particle size analysis of a soil sample involves determining the percentage by mass of particles within the different size ranges. Particle sizes in soils can vary from less than 0.001mm to over 100mm. In China, British and Japan Standards, the size ranges detailed in Fig. 1.4.1 are specified. The same terms are also used to describe particular types of soil. The particle size distribution of a coarse-grained

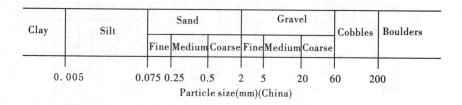

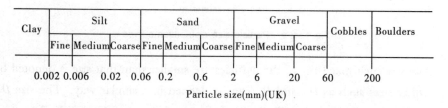

Fig. 1.4.1 Particle size ranges

soil can be determined by the method of sieving. The particle size distribution of a fine-grained soil or the fine grained fraction of a coarse-grained soil can be determined by the method of sedimentation.

(1) Method of sieving

The soil sample is passed through a series of standard test sieves having successively smaller mesh sizes. The mass of soil retained in each sieve is determined and the cumulative percentage by mass passing each sieve is calculated. The particle size distribution of a soil is presented as a curve on a semilogarithmic plot. The ordinate is the percentage by mass of particles smaller than the size given by the abscissa. Examples of particle size distribution curves appear in Fig. 1.4.2. The particle size corresponding to any specified value on the "percentage smaller" scale can be read from the particle size distribution curve.

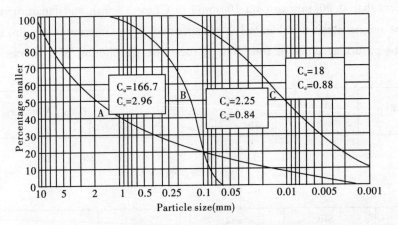

Fig. 1.4.2 Particle size distribution curves

The size such that 10% of the particles are smaller than that size is denoted by D_{10}. Other sizes such as D_{30} and D_{60} can be defined in a similar way. The size D_{10} is defined as the effective size. The general slope and shape of the distribution curve can be described by means of the coefficient of uniformity (C_u) and the coefficient of curvature (C_c), defined as follows:

$$C_u = \frac{D_{60}}{D_{10}} \tag{1.4.1}$$

$$C_c = \frac{D_{30}^2}{D_{60} D_{10}} \tag{1.4.2}$$

The higher the C_u value, the larger the range of particle sizes. A well-graded soil has $1 < C_c < 3$.

(2) Method of sedimentation

This method is based on Stokes' law which governs the velocity at which spherical particles settle in a suspension: the larger is the particle, the greater is the settling velocity and vice versa.

1.5 Density properties for granular soils

Cohesionless soils include stone, crushed stone, gravel and sand soil, which all present single grain fabric with no cohesive force and no cementation effect. Compactness is the main factor affecting the engineering properties in terms of this kind of soil. Hence, if the soil is more compact, the bearing capacity will be higher, the compressibility will be smaller, and then the stability will be much better accordingly. Since the soil particles are coarse and relative surface area is quite small, water has little influence on engineering properties in terms of this kind of soil. There is only strong bound water on the surface, and no weakly bound water.

How to express the compactness of granular soils is the key point. Besides the coefficient of uniformity C_u, there are usually some other ways.

1.5.1 Void ratio

Figure 1.5.1 shows two of the many possible ways that a system of equal-sized spheres can be packed. The dense packings represent the densest possible state for such a system. Looser systems than the simple cubic packing can be obtained by carefully constructing arches within the packing, but the simple cubic packing is the loosest of the stable arrangements. The void ratio and porosity of these simple packings can be computed from the geometry of the packings, and the results are given in Table 1.5.1.

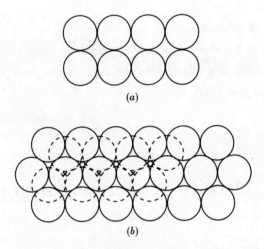

Fig. 1.5.1 Arrangement of uniform spheres

(a) Plan and elevation view: simple cubic packing. (b) Plan view: dense packing. Solid circles, first layer; dashed circles, second layer; ○, location of sphere centers in third layer; face-centered cubic array; ×, location of sphere centers in third layer; close-packed hexagonal array. (From Deresiewicz, 1958)

Table 1.5.1 Maximum and Minimum Densities for Granular Soils

Description	Void Ratio		Porosity		Dry Unit Weight (pcf)	
	e_{max}	e_{min}	n_{max}	n_{min}	γ_{dmax}	γ_{dmin}
Uniform spheres	0.92	0.35	47.6	26.0	—	—
Standard Ottawa sand	0.80	0.50	44	33	92	110
Clean uniform sand	1.0	0.40	50	29	83	118
Uniform inorganic silt	1.1	0.40	52	29	80	118
Silty sand	0.90	0.30	47	23	87	127
Fine to coarse sand	0.95	0.20	49	17	85	138
Micaceous sand	1.2	0.40	55	29	76	120
Silty sand and gravel	0.85	0.14	46	12	89	146

B. K. HOUGH, *Basic Soils Engineering*.

The smaller the range of particles sizes present (i. e., the more nearly uniform the soil), and the more angular the particles present, the smaller the minimum density (i. e., the greater the opportunity for building a loose arrangement of particles). The greater the range of particle sizes present, the greater the maximum density (i. e., the voids among the larger particles can be filled with smaller particles).

According to the engineering experience, Table 1.5.2 characterizes the density of granular soils in the basis of void ratio.

Table 1.5.2 Division of compactness of sandy soils in terms of void ratio

Soil type	Compactness			
	Dense	Medium dense	Less dense	Loose
Gravel, coarse or medium sand	$e<0.6$	$0.60 \leqslant e \leqslant 0.75$	$0.75 < e \leqslant 0.85$	$e > 0.85$
Fine sand, silt	$e<0.7$	$0.70 \leqslant e \leqslant 0.85$	$0.85 < e \leqslant 0.95$	$e > 0.95$

Values of water content for natural granular soils vary from less than 0.1% for air-dry sands to more than 40% for saturated, loose sand.

1.5.2 Relative density

A useful way to characterize the density of a natural granular soil is with relative density D_r, defined as

$$D_r = \frac{e_{max} - e}{e_{max} - e_{min}} \times 100\% = \frac{\gamma_{dmax}}{\gamma_d} \times \frac{\gamma_d - \gamma_{dmin}}{\gamma_{dmax} - \gamma_{dmin}} \times 100\% \qquad (1.5.1)$$

where

e_{min} = void ratio of soil in densest condition

e_{max} = void ratio of soil in loosest condition

e = in-place void ratio

γ_{dmax} = dry unit weight of soil in densest condition

γ_{dmin} = dry unit weight of soil in loosest condition

γ_d = in-place dry unit weight

According to the engineering experience, divide the compactness of sandy soils in terms of relative density D_r:

$$0 < D_r \leq 0.33 \quad \text{loose}$$
$$0.33 < D_r \leq 0.67 \quad \text{medium dense}$$
$$0.67 < D_r \leq 1.0 \quad \text{dense}$$

A variety of tests have been proposed to measure the maximum and minimum void ratios (Kolbuszewski, 1948). The test to determine the maximum density usually involves some form of vibration. The test to determine minimum density usually involves pouring oven-dried soil into a container. Unfortunately, the details of these tests have not been entirely standardized, and values of the maximum density and minimum density for a given granular soil depend on the procedure used to determine them. By using special measures, one can obtain densities greater than the so-called maximum density. Densities considerably less than the so-called minimum density can be obtained, especially with very fine sands and silts, by slowly sedimenting the soil into water or by fluffing the soil with just a little moisture present.

1.5.3 Standard penetration test

Using void ratio e and relative density D_r to express the compactness of soils has many defects. Besides a lot of experimental influencing factors, it is very hard to get the intact specimens of the granular soils. Thus many field experiments have been proposed (i.e. standard penetration test, the cone penetration test) to get the compactness of soils.

According to *Code for Design of Building Foundation* (GB 50007 - 2011), Table 1.5.3 presents a correlation of standard penetration resistance with relative density for sand.

Table 1.5.3 Standard penetration test

Relative Density	Dense	Medium dense	Less dense	Loose
Penetration resistance N	$30 < N$	$15 < N \leq 30$	$10 < N \leq 15$	$N \leq 10$

The standard penetration test is a very valuable method of soil investigation. It should, however, be used only as a guide, because there are many reasons why the results are only approximate.

1.6 Plasticity properties of soils

1.6.1 Plastic behavior and consistency limits

Plasticity is a very important characteristic of fine-grained soil. In general, depending on its water content, a soil may exist in one of the liquid, plastic, semi-solid and solid states. If the water content of a soil initially in the liquid state is gradually reduced, the state will change from liquid through plastic and semi-solid, accompanied by gradually reducing volume, until the solid state is reached. The water content at the division between the solid and the semi-solid state is the shrinkage limit w_s. The division between the semi-solid and plastic state is the plastic limit w_p. The water content indicating the division between the plastic and liquid state has been designated the liquid limit w_L. The liquid limit, plastic limit and shrinkage limit are also called consistency limits (see Fig. 1.6.1).

The liquid and plastic limits of the range of water content are defined as the plasticity index I_p, i. e.

$$I_p = w_L - w_p \qquad (1.6.1)$$

For proper evaluation of a soil's plasticity properties, it has been found desirable to use both the liquid limit and plasticity index values. The natural water content (w) of a soil relative to the liquid and plastic limits can be represented by means of the liquidity index (I_L), i. e.

$$I_L = \frac{w - w_p}{w_L - w_p} \qquad (1.6.2)$$

A very low value for the I_L, or a value near zero, indicates that the water content is near the plastic limit, where experience has shown that the sensitivity will be low and the cohesive strength relatively high. As the natural water content approaches the liquid limit, the sensitivity increases.

The liquid and plastic limits are determined by means of arbitrary test procedures.

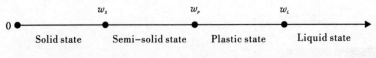

Fig. 1.6.1   Consistency limits

1.6.2   Liquid and plastic limit tests

(1) Determination of the plastic limit

There are two methods to determine the plastic limit: Rubbing method and liquid-plastic limit combined device method.

The rubbing method is relatively simple to carry out; the apparatus itself is simple, too. First, we need to make several specimens. Then, take a specimen and roll it into a ball and then we put it on a ground glass sheet to roll it evenly into a thread of soil with our palms until the diameter of the thread reaches 3mm, Fig. 1.6.2. This procedure of rolling continues until the thread starts to crumble or there is no shadow mark on the glass just as the diameter of 3mm is reached; at this point the water content of the specimen is determined. The same process is carried out in other specimens and the average water content is stated as the plastic limit of the soil. In spite of the seeming arbitrary nature of this test procedure, an experienced technician can obtain a reasonable plastic limit results for the engineering construction.

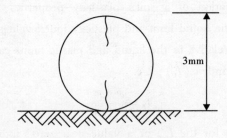

Fig. 1.6.2   Rubbing method

Liquid-plastic combine tester is shown in Fig. 1.6.3. First, we make three kinds of specimens with different water content. Make the three kinds of well-prepared specimens fully mixed and put them in the soil cup. Then with the electromagnetic fall-cone test, the cone is released to penetrate the soil paste for exactly 5s and mark down its depth, introduced in *Standard for soil test method* (GB/T 50123 −1999). The penetration procedure is repeated three times on the specimens. A log-log plot is drawn of water content/penetration depth and three points should be in a line in this plot. So the water content corresponding to a penetration of 2mm is the value of plasticity limit.

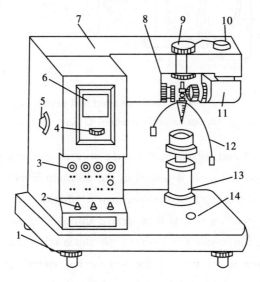

Fig. 1.6.3　Liquid – Plastic limit combined device

1. Horizontal adjustment screw; 2. Pilot switch; 3. Indicate lamp; 4. Nil-line adjustment screw;
5. Retroreflector adjustment screw; 6. Screen; 7. Chassis; 8. Objective lens adjustment screw;
9. Electromagnetic apparatus; 10. Luminous source adjustment screw; 11. Luminous source;
12. Cone apparatus; 13. Elevator platform; 14. Spirit bubble

(2) Determination of the liquid limit

There are two methods for the liquid limit test: liquid-plastic limit combined device method and Casagrande method, namely dish-type liquid limit device method.

Although the liquid-plastic limit combined device method is preferred, the Casagrande method of determining liquid limit is still widely used. The apparatus used is shown in Fig. 1.6.4 and consists basically of a metal dish which may be raised by rotating a cam and then allowed to fall through at a height of 10mm on to a hard rubber block.

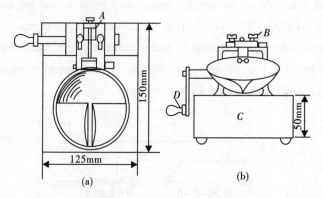

Fig. 1.6.4 Dish-type liquid limit device
A-Adjustment plate; B-Screw; C-Base; D-Handle

The soil is dried and mixed in the same way as in the previously described method. Some of the soil paste is placed in the dish and leveled-off parallelly with the rubber base until the depth of the specimen is 10mm.

The standard grooving tool is then drawn through the soil paste to form the groove. By turning the handle (at 2 rvs/s) the cup is raised and dropped on to the rubber base until the lower part of the groove has closed up over a length of 13mm. The number of blows (number of revolutions) required for this is recorded.

The dish is then refilled with the same paste mixture and the grove-closing procedure is repeated several times and an average number of blows are obtained for that mixture. After this, a small portion of the paste is taken and its water content is found. The whole procedure is then repeated with paste mixture having different water contents for five or six times in all.

1.6.3 Two important indexes and the engineering application

(1) Plasticity index I_p

It can be seen from Fig. 1.6.1 that if the value of plasticity limit w_L and liquid limit w_p has big difference, the range of plasticity of the soil will be wider, the weakly bound water on the surfaces of soil particles will be thicker, the specific surface area of the soil particles will be bigger and the surface absorption capability of soil particles will be greater. Thus, plasticity index, to a large extent, indicates the amount of clay particles, colloidal particles and clay mineral contents of the soil.

According to *Code for design of building foundation* (GB 50007 −2011) and *Code for rock and soil classification of railway engineering* (TB 10077 − 2001), divide the cohesive soils on the basis of the range of plasticity index I_p.

$10 < I_p \leqslant 17$ Silty clay

$I_p > 17$ Clay

(2) Liquidity index I_L

From the Eq. (1.6.2) and Fig. 1.6.1, it can be seen that if the natural water content is less than the plastic limit ($w \leqslant w_p$), liquidity index will be negative ($I_L \leqslant 0$) and the soil will be in solid state; if the natural water content is greater than the liquid limit ($w \geqslant w_L$), liquidity index will be equal or greater than 1.0 ($I_L \geqslant 1.0$) and the soil will be in flow condition. Thus, liquidity index indicates the degree of hardness of cohesive soil.

According to *Code for investigation of geotechnical engineering* (GB 50021 − 2001) and *Code for design of building foundation* (GB 50007 −2011), divide the cohesive soils into five states on the basis of liquidity index I_L, shown in Table 1.6.1.

According to *Code for rock and soil classification of railway engineering* (TB 10077 −2001), divide the cohesive soils into four states on the basis of liquidity index I_L, shown in Table 1.6.2.

Table 1.6.1 Division of hardness of cohesive soil

Degree of hardness	Solid	Stiff-plastic	Malleable	Soft-plastic	Flowing
Liquidity index	$I_L \leq 0$	$0 < I_L \leq 0.25$	$0.25 < I_L \leq 0.75$	$0.75 < I_L \leq 1.0$	$1.0 < I_L$

Table 1.6.2 Division of hardness of cohesive soil

Degree of hardness	Solid	Stiff-plastic	Soft-plastic	Flowing
Liquidity index	$I_L \leq 0$	$0 < I_L \leq 0.5$	$0.5 < I_L \leq 1.0$	$1.0 < I_L$

1.7 Soil compaction

Compaction is the process of increasing the density of a soil by packing the particles closer together with the reduction in the volume of air: there is no significant change in the volume of water of the soil. In the construction of embankments, loose soil is placed in layers ranging from 75mm to 450mm in thickness, each layer being compacted to a specified standard by means of rollers, vibrators or rammers. In general, the higher the degree of compaction, the higher will be the strength and the lower will be the compressibility of the soil.

The degree of compaction of a soil is measured in terms of dry density, i.e. the mass of the solid only per unit volume of soil. If the bulk density of the soil is ρ and the water content is w, then it is apparent that the dry density is given by:

$$\rho_d = \frac{\rho}{1+w} \qquad (1.7.1)$$

The compaction characteristics of a soil can be assessed by means of the standard compaction test. The soil is compacted in a cylindrical mould using a standard compactive effort. In the test the volume of the mould is 947mm^3 and the soil (with all particles larger than 5mm removed) is compacted by a rammer consisting of a 2.5kg mass falling freely through 305mm: the soil is compacted in three equal layers, each layer receiving 25 blows with the rammer. In the modified

A. A. S. H. O. test the mould is the same as used in the above test, but the rammer consists of a 4.5kg mass falling though 457mm; the soil (with all particles larger than 40mm removed) is compacted in five layers, each layer receiving 56 blows with the rammer.

The effectiveness of the compaction process is dependent on several factors:

(1) The water content of soil. The maximum dry density can't be obtained if there is just a little or a lot of water in the soil. If the water content is quite low, there is basically strong bound water in the soil and the bound water film is too thin, adding the influence of interparticle friction and attraction, which make the soil particles not quite easy to move, thus not easy to get compact. If the water content is quite large, there is relatively a lot of free water in the soil, which is considered incompressible under the engineering loads. Since the free water take up a certain space, the soil with large water content is not easy to get compact, either. When the water content of soil is just the optimal water content, there is some weakly bound water and no free water in the soil. Weakly bound water film adhere to the soil particles and can move together with the soil particles, during which weakly bound water film has the lubrication effect, making soil particles easy to move, fill the voids and compact. Thus the maximum dry density can be obtained.

(2) The energy supplied by the compaction equipment (referred to as the compactive effort).

(3) The nature and type of soil (i. e. sand or clay; uniform or well graded, plastic or non-plastic).

(4) The large particles in the soil.

After compaction using one of the two standard methods, the bulk density and water content of the soil are determined and the dry density is calculated. For a given soil the process is repeated at least five times, the water content of the sample being increased each time. Dry density is plotted against water content and a curve of the form shown in Fig. 1.7.1 is obtained. This curve shows that for a particular method of compaction (i. e. a particular compactive effort) there is a particular value of water content, known as the optimum water content (w_{op}), at which a maximum value of dry density is obtained. At low values of water content, most soils tend to be stiff and are difficult to compact. As the water content increases, the soil becomes more workable, facilitating compaction and resulting in higher dry densities, which is very beneficial for engineering construction. At high water contents, however, the dry density decreases with increasing water content, and an increasing proportion of the soil volume is occupied by water. So, dry density is quite an important quality inspection index for compaction.

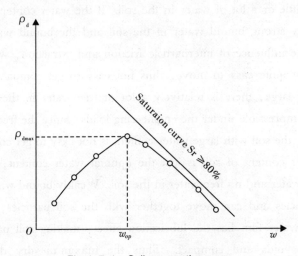

Fig. 1.7.1 Soil compaction curve

1.8 Engineering classification of soils

The direct approach to the solution of a soil engineering problem consists of first measuring the soil property needed and then employing this measured value in some rational expression to determine the answer to the problem.

To measure fundamental soil properties like permeability, compressibility, and strength can be difficult, time consuming, and expensive. In many soil engineering problems, such as pavement design, there are no rational expressions available for the analysis for the solution. For these reasons, sorting soils into groups showing similar behavior may be very helpful. Such sorting is soil classification.

In our country, several kinds of classification standards pertain: *Standard for Engineering Classification of Soil* (GB/T 50145 −2007); *Code for Design of Building Foundation* (GB 50007 −2011); *Test Methods of Soils for Highway Engineering* (JTG E40 −2007); *Specification of Soil Test* (SL 237 −1999). The basic classification of soil in building foundation will be introduced as follows.

1.8.1 Crushed stones, gravels

It refers to the soils in which the particles (size >2mm) take up more than 50% of the whole mass weight. The particle size and gradation of this kind of soil have great impact on engineering construction. Thus according to particle size and gradation, classify this kind of soil shown in Table 1.8.1.

Table 1.8.1 Classification of crushed stones and gravels

Soil type	Soil particle shape	Gradation
Boulder	Round forms dominant	Particles (size >200mm) take up more than 50% of the whole mass weight.
Block stone	Angular forms dominant	
Cobble	Round forms dominant	Particles (size >20mm) take up more than 50% of the whole mass weight.
Crushed stone	Angular forms dominant	
Round gravel	Round forms dominant	Particles (size >2mm) take up more than 50% of the whole mass weight.
Angular gravel	Angular forms dominant	

1.8.2 Sand

It refers to the soils of which the particles from size >2mm take up less than 50% of the whole mass weight to size >0.075mm take up more than 50% of the whole mass weight.

For this kind of soil, the classification code is similar to the crushed stones and gravels, shown in Table 1.8.2.

Table 1.8.2　Classification of sand in the basis of gradation

Soil type	Gradation
Gravelly sand	Particles (size >2mm) take up 25% −50% of the whole mass weight.
Coarse sand	Particles (size >0.5mm) take up 50% of the whole mass weight.
Medium sand	Particles (size >0.25mm) take up 50% of the whole mass weight.
Fine sand	Particles (size >0.075mm) take up 85% of the whole mass weight.
Silty sand	Particles (size >0.075mm) take up 50% of the whole mass weight.

According to the void ratio, relative density and standard penetration test, this kind of soil can also be classified, shown in section 1.5.

1.8.3　Silty soil

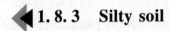

It refers to the soils in which particles (size >0.075mm) take up less than 50% of the whole mass weight and the plasticity index $I_p \leqslant 10$.

1.8.4　Clay

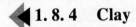

In the basis of plasticity index I_p, clay can be divided as shown in section 1.6.3.

On the basis of the degree of hardness, clay can be divided as shown in Table 1.6.1 and Table 1.6.2.

EXERCISES

1.1 Explain the concept of plasticity index and liquidity index and its application in engineering.

1.2 What is particle size distribution curve and how to use it in engineering?

1.3 How to appraise the compactness of granular soils?

1.4 What are maximum dry density and optimum water content of soil?

1.5 A saturated clay specimen has the water content of 36.0%, and particle specific gravity of 2.70. Determine the void ratio and dry density.

1.6 The volume of a soil specimen is 60cm³, and its mass is 108g. After being dried, the mass of the sample is 96.43g. The value of d_s is 2.7. Calculate wet density, dry density, water content, porosity and the degree of saturation.

1.7 A soil specimen has a water content w_1 of 12% and unit weight γ_1 of 19.0kN/m^3. Keep void ratio constant and increase water content to $w_2 = 22\%$. How much water will be infused into 1m^3 soil?

1.8 There is a sample which is got from a certain natural sand column. The water content measured by experiment is 11%, bulk density $\rho = 1.70$g/cm^3, the minimum dry density is 1.41g/cm^3, the maximum dry density is 1.75g/cm^3. Determine the sandy soil's dense degree.

1.9 Calculate the dry unit weight, the saturated unit weight and the buoyant unit weight of a soil having a void ratio of 0.70 and a value of d_s of 2.72. Also calculate the unit weight and water content at a degree of saturation of 75%.

REFERENCES

1. Karl Terzaghi (1943). Theoretical Soil Mechanics[M]. John Wiley and Sons, New York.
2. T. William Lambe and Robert V. Whitman (1969). Soil Mechanics[M]. John Wiley and Sons, New York.
3. John Atkinson (1993). An Introduction to the Mechanics of Soils and Foundations[M]. McGraw-Hill, England.
4. R. F. Craig (1998). Soil Mechanics[M]. E and FN Spon, London and New York.
5. Soil mechanics work team at Hohai University (2004). Soil Mechanics[M]. China Communication Press, Beijing.
6. Shude Zhao, Hongjian Liao (2010). Soil Mechanics (The Second Edition) [M]. Higher Education Press, Beijing.
7. Shude Zhao, Hongjian Liao(2009). Civil Engineering Geology [M]. Science Press, Beijing.
8. Ministry of Housing and Urban-Rural Construction of the People's Republic of China (2012), GB 50007 −2011 Code for Design of Building Foundation[S], China Building Industry Press, Beijing.
9. Ministry of Water Resources of the People's Republic of China (2000), GB/T 50123 −1999 Standard for soil test method[S], China Planning Press, Beijing.

10. Ministry of Construction of the People's Republic of China (2009), GB 50021—2001 Code for investigation of geotechnical engineering (The 2009 Revised Edition) [S], China Building Industry Press, Beijing.
11. Ministry of Railways of the People's Republic of China (2001), TB 10077—2001 Code for Rock and Soil Classification of Railway Engineering[S], China Railway Press, Beijing.
12. Ministry of Water Resources of the People's Republic of China (2008), GB/T 50145—2007 Standard for Engineering Classification of Soil [S], China Planning Press, Beijing.
13. Ministry of Communications of the People's Republic of China (2007), JTG E40—2007 Test Methods of Soils for Highway Engineering [S], China Communications Press, Beijing.
14. Ministry of Water Resources of the People's Republic of China (2003), SL 237—1999 Specification of Soil Test[S], China Standards Press, Beijing.

SOIL MECHANICS CHAPTER 2

Permeability of soils and seepage force

2.1 Introduction

Any given mass of soil consists of solid particles of various sizes and interconnected void spaces among them. When the soil is directly used as the material of hydraulic structures, water can flow by the effect of water level difference from one side of high energy to another side of low energy through the continuous voids in a soil, which is quite common in the slope, earth dam, groundsill and foundation pit, and so on.

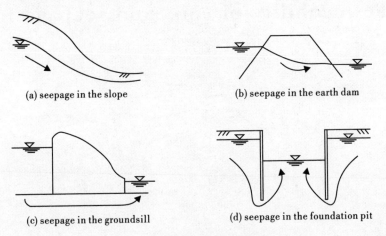

(a) seepage in the slope (b) seepage in the earth dam

(c) seepage in the groundsill (d) seepage in the foundation pit

Fig. 2.1.1 Seepage phenomena

As the above-mentioned effect of water level difference, permeability is defined as the property of a soil that allows the seepage of fluids through its interconnected void spaces.

The water seepage in a soil can make complex interactions between water and soil, which generate various problems in the engineering construction. On one hand, with the change of amount, quality, storage location of water itself, a lot of problems come about. For examples, the seepage causes head loss in the water retaining structures (such as floodgate and dam) and the infiltration of sea water or waste water leads to groundwater pollution. On the other hand, water seepage can greatly change the stress state in soil, which significantly influences the consolidation and strength of soil mass, and thus results in the change of stable condition or even destruction accidents in the hydraulic structures or foundation soil.

It can be seen from the above examples that discussing the influence of the fluid on the soil through which it is flowing is significant. This chapter is devoted to the study of the basic parameters and relevant engineering problems involved in the flow of water through soils.

 2.2 Capillary phenomena

There are many evidences that a liquid surface resists tensile forces because of the attraction between adjacent molecules in the surface. This attraction is measured by surface tension, a constant property of any pure liquid in contact with another given liquid or with a gas at a given temperature. An example of these evidences is the fact that in fine-grained soils (the average diameter of voids is 0.002 −0.5mm) water is capable of rising to a considerable height above the water table and remaining there permanently. This kind of water is called capillary water. This phenomenon is commonly referred to as capillarity.

The soil layer wetted by capillarity is called capillary water zone. Based on the formation and distribution conditions, capillary water zone can be divided into 3 parts. They are normal capillary water zone, capillary network zone and capillary suspension zone respectively.

(1) Normal capillary zone (zone of capillary saturation). This part lies at the bottom of the capillary water zone. The capillary water is mainly formed with the water table rising directly. All the voids are almost filled with capillary water. The

normal capillary zone accordingly makes change with the rise and fall of underground water level.

(2) Capillary network zone. This part lies in the middle of the capillary water zone. With the quick fall of groundwater, this part also dip quickly accordingly, but some capillary water in the thinner capillary voids can't make quick movement and still stays in the voids; with the capillary water drops in some larger voids, air bubbles are left in the voids, making capillary water presents netted distribution. The water particles in this part can move with the effect of surface tension and gravity.

(3) Capillary suspension zone. This part lies on the top of the whole capillary water zone. The capillary water in this part is formed by the infiltration of surface water. The water particles are suspended in the soil particles and have no connection with the other capillary water in the middle and bottom region. This part of suspended capillary water can make downward movement with the effect of gravity by atmosphere precipitation.

The above-mentioned three kinds of capillary water zones don't necessarily exist at the same time, which depends on the local hydrogeological conditions. When the underground water level is high, maybe only normal capillary zone exists; on the contrary, when the underground water level is low, the three kinds of capillary water can exist at the same time.

The capillary water in soil has great impact on engineering constructions, most of which are listed below.

(1) Since capillary water is above the surface of the free water, it leads to the change of dry and wet states in roadbed. Hence, the rise of capillary water is one of the main factors resulting in the roadbed freeze-harm;

(2) For buildings, the rise of capillary water can cause too much moisture in the basement, so the damp course becomes an absolute necessity at the top of the basement;

(3) The rise of capillary water can lead to the soil paludification and salinization, affecting engineering construction and agricultural economic development.

However, with the use of the temporary cohesion generated by the capillary phenomena in the wet sand, there exists the upright slope of 2—3m height and sand sculpture art in the coastal region.

2.3 Darcy's law

2.3.1 Laminar flow and turbulent flow

(1) Laminar Flow

In the groundwater seepage, the flow lines formed by the water particles are paralleled in all places. The soil located on the bottom can not be floated up; the leaves or something light floating on the water surface are still there and can not be pulled down; through a certain space, water flows smoothly with uniform velocity and the flow velocity in the midst of water cross section is higher, on both sides of which is smaller. With all the flow characteristics stated above, the flow is commonly referred to as laminar flow. The flow velocity, flow direction, water level, hydraulic pressure and some other motion indexes at any point in this seepage field don't change over time, which is called steady flow movement.

(2) Turbulent Flow

Compared with laminar flow, in this kind of groundwater seepage, the flow lines cross each other. The flow presents twisted, mixed and irregular movement and exists hydraulic drop and whirlpools. With all the flow characteristics stated above, this flow is commonly referred to as turbulent flow. All the motion indexes at any point in this seepage field change over time, which is called unsteady flow movement.

2.3.2 Darcy's law

The voids in a soil (sand, clay) are generally quite small. Although the real seepage of water through the small void spaces in a soil is irregular, the flow can be regarded as laminar flow, because that the movement of water through the void spaces is very slow. In the condition of laminar flow, the discharge velocity and energy loss obeys the linear seepage relationship, which was obtained by H. Darcy in 1856.

In order to obtain a fundamental relation for the quantity of seepage through a soil mass under a given condition, the case shown in Fig. 2.3.1 is considered. The cross-sectional area of the soil is equal to A and the rate of seepage is Q. Darcy experimentally found that Q was proportional to i, and that

$$Q = k \cdot A \cdot i \qquad (2.3.1)$$

Note that A is the cross sectional area of the soil perpendicular to the direction of flow.

The hydraulic gradient i can be given by

$$i = \frac{H_1 - H_2}{L} = \frac{\Delta H}{L}$$

where L is the distance between the two piezometric tube.

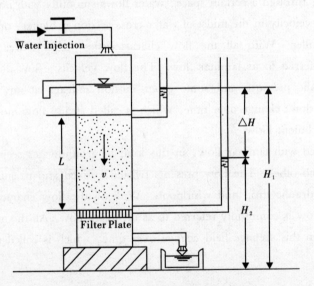

Fig. 2.3.1 Permeability test device

A further consideration of the velocity at which a drop of water moves as it flows through soil helps to understand fluid flow. Equation (2.3.1) can be rewritten as

Chapter 2 Permeability of soils and seepage force

$$\frac{Q}{A} = k \cdot i = v$$

Therefore, for sandy soils, Darcy published a linear relation between the discharge velocity and the hydraulic gradient:

$$v = ki \qquad (2.3.2)$$

where v = discharge velocity, cm/s, which is not the actual velocity of seepage through soil, the velocity v given by Eq. (2.3.2) is the discharge velocity calculated on the basis of the gross cross-sectional area.

i = hydraulic gradient, the water level difference per unit length along the flow direction.

k = coefficient of permeability, cm/s, a measure of the resistance of the soil to flow of water.

Darcy's law given by Eq. (2.3.2), $v = ki$, is true for laminar flow through the void spaces. Several studies have been conducted to investigate the range over which Darcy's law is valid, and an excellent summary of these works was given by the Reynolds number. For flow through soils, Reynolds number R_n can be given by the relation

$$R_n = \frac{vD\rho}{\mu} \qquad (2.3.3)$$

where v = discharge velocity, cm/s

D = average diameter of the soil particle, cm

ρ = density of the fluid, g/cm^3

μ = coefficient of viscosity, g/(cm · s)

For laminar flow conditions in soils, experimental results showed that

$$R_n = \frac{vD\rho}{\mu} \leqslant 1$$

With coarse sand, assuming $D = 0.45$m and $k \approx 100D^2 = 100(0.045)^2 = 0.203$cm/s,

Assuming $i = 1$, then $v = ki = 0.203$cm/s. Also, $\rho_{water} \approx 1$g/cm^3, and $\mu_{20℃} = (10^{-5})(981)$g/(cm · s). Hence

$$R_n = \frac{(0.203)(0.045)(1)}{(10^{-5})(981)} = 0.931 < 1$$

From the above calculations, we can conclude that, for flow of water through all types of soil (sand, silt, and clay), the flow is laminar and Darcy's law is valid. With coarse sand, gravels, and boulders, turbulent flow of water can be expected.

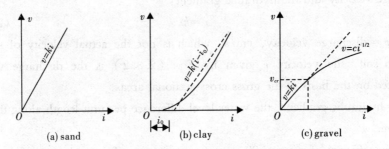

Fig. 2.3.2 Discharge velocity—hydraulic gradient relationship of soil

Darcy's law as defined by Eq. (2.3.2) implies that the discharge velocity bears a linear relation with the hydraulic gradient for sand, as shown in Fig. 2.3.2 (a).

Hansbo (1960) reported the results of four undisturbed natural clays, as shown in Fig. 2.3.2 (b). On the basis of his results

$$v = k(i - i_0) \qquad i \geq i_0$$

and

$$v = ki^n \qquad i < i_0$$

The value of n for the four Swedish clays was about 1.6. There are several studies, however, that refute the preceding conclusion.

For gravel or some other coarse grained soils, the discharge velocity bears a linear relation with the hydraulic gradient just when the hydraulic gradient is small; when the hydraulic gradient is quite large, turbulent flow of water can be expected. Discharge velocity and hydraulic gradient will not present linear relationship, as shown in Fig. 2.3.2(c).

It must be pointed out that the velocity v given by Eq. (2.3.2) is the discharge velocity calculated on the basis of the gross cross-sectional area. Since water can flow only through the interconnected pore spaces, the actual velocity of seepage through soil is v'. Assuming the rate of seepage is Q, the cross-sectional area of the soil is equal to A. Hence the actual cross-water area

$$A' = nA$$

According to the continuity of water, $Q = vA = v'A'$, then

$$v = v' \times \frac{A'}{A} = v'n = v'\frac{e}{1+e} \qquad (2.3.4)$$

where e is the void ratio of the soil, n is the porosity of the soil.

2.4 Determination of permeability coefficient

2.4.1 Laboratory methods

The coefficient of permeability k, a soil parameter, reflects the seepage ability of a soil mass. It can be interpreted as the superficial velocity for a gradient of unity, i.e., $k = v$ for a gradient equal to 1. Thus, the value of k is an important measurement criteria for the soil seepage ability. It can't be calculated directly and needs to be measured in laboratory tests.

The two most common laboratory methods for determining the coefficient of permeability of soils are constant-head test and falling-head test. In both cases water flows through a soil sample and the rates of flow and the hydraulic gradients are measured.

It needs to be noted that values of the coefficient of permeability measured in laboratory permeameter tests are often highly inaccurate, for a variety of reasons such as anisotropy (i. e. values of k different for horizontal and vertical flow) and small samples being unrepresentative of large volumes of soil in the ground, and in practice values of k measured from in situ tests are much better.

(1) Constant-Head Test

The constant-head test is suitable for more permeable granular materials. The basic laboratory test arrangement is shown in Fig. 2.4.1. The soil specimen is placed inside a cylindrical mold, and the constant-head loss h of water flowing through the soil is maintained by adjusting the supply. The outflow water is collected in a measuring cylinder, and the duration of the collection period is recorded. From Darcy's law, the total quantity of flow Q in time t can be given by

$$Q = vAt = kiAt$$

where A is the area of cross section of the specimen. However, $i = h/L$, where L is the length of the specimen, so $Q = k(h/L)At$. Rearranging gives

$$k = \frac{QL}{Aht} \qquad (2.4.1)$$

Once all the quantities on the right-hand side of Eq. (2.4.1) have been determined from the test, the coefficient of permeability of the soil can be calculated.

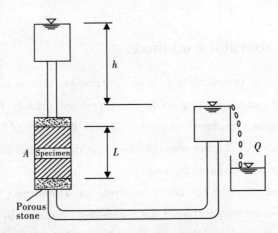

Fig. 2.4.1 Constant-head laboratory permeability test

(2) Falling-Head Test

The falling-head permeability test is more suitable for fine-grained soils. Figure 2.4.2 shows the general laboratory arrangement for the test. The soil specimen is placed inside a tube, and a standpipe is attached to the top of the specimen. Water that comes from the standpipe flows through the specimen. The initial head difference h_1 at time $t = t_1$ is recorded, and then water is allowed to flow through the soil such that the final head difference at time $t = t_2$ is h_2.

The rate of flow through the soil is

$$q = kiA = k\frac{h}{L}A = -a\frac{dh}{dt} \qquad (2.4.2)$$

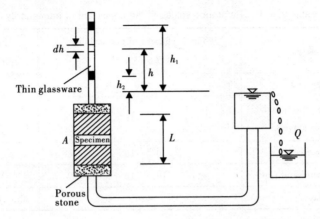

Fig. 2.4.2 Falling-head laboratory permeability test

where h = head difference at any time t
A = area of specimen
a = area of standpipe
L = length of specimen

From Eq. (2.4.2),

$$k = \frac{aL}{A(t_2 - t_1)} \ln \frac{h_1}{h_2} \qquad (2.4.3a)$$

or

$$k = 2.3 \frac{aL}{A(t_2 - t_1)} \lg \frac{h_1}{h_2} \qquad (2.4.3b)$$

The values of a, L, A, t_1, t_2, h_1 and h_2 can be determined from the test, and the coefficient of the permeability k for a soil can then be calculated from Eq. (2.4.3a) or Eq. (2.4.3b).

There are several factors affecting the coefficient of permeability, such as soil types, gradation, void ratio, water temperature, and so forth. Hence, in order to make precise measurement of the coefficient of permeability, we must try to keep the original state of a soil and eliminate any possible influences of artificial interference. The reference values of the coefficient of permeability of some soil types are listed in Table 2.4.1.

Table 2.4.1   Reference values of the coefficient of permeability

Basic soil type	The coefficient of permeability $k/(\text{cm} \cdot \text{s}^{-1})$	The degree of permeability
Pure gravels	$>10^{-1}$	High
Mix of pure gravels and some other gravels	$10^{-3} \sim 10^{-1}$	Middle
The finest sand	$10^{-5} \sim 10^{-3}$	Low
Mix of silt, sand and clay	$10^{-7} \sim 10^{-5}$	Very low
Clay	$<10^{-7}$	Almost impermeable

◀ 2.4.2   Effective coefficient of permeability for stratified soils

In general, natural soil deposits are stratified. If the stratification is continuous, the effective coefficients of permeability for flow in the horizontal and vertical directions can be readily calculated.

(1) Flow in the horizontal direction

Figure 2.4.3 shows several layers of soil with horizontal stratification. Owing to fabric anisotropy, the coefficient of permeability of each soil layer may vary depending on the direction of flow. Therefore, let us assume that $k_1, k_2, \cdots, k_n$ are the coefficient of permeability for layers $1, 2, \cdots, n$, respectively, for flow in the horizontal direction.

Considering unit width of the soil layers as shown in Fig. 2.4.3, the rate of seepage in the horizontal direction can be given by

$$Q_x = Q_{1x} + Q_{2x} + \cdots + Q_{nx} = \sum_{i=1}^{n} Q_{ix} \qquad (a)$$

where Q is the flow rate through the combined stratified soil layers and $Q_{1x}, Q_{2x}, \cdots, Q_{nx}$ is the rate of flow through soil layers $1, 2, \cdots, n$, respectively. Note that for flow in the horizontal direction (which is the direction of stratification of the soil layers), the hydraulic gradient is the same for all layers ($i_1 = i_2 = \cdots i_n = i = \frac{\Delta h}{L}$). So

Chapter 2 Permeability of soils and seepage force

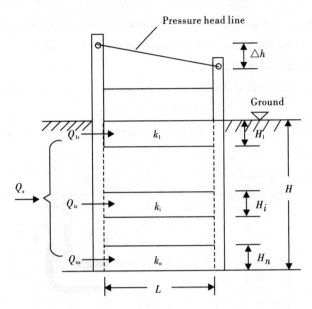

Fig. 2.4.3 Flow in horizontal direction in stratified soil

$$Q_{1x} = k_1 i H_1$$
$$Q_{2x} = k_2 i H_2$$
$$\cdots$$
$$Q_{nx} = k_n i H_n \tag{b}$$

where i = hydraulic gradient

$H_1, H_2, \ldots H_n$ = thicknesses of layers 1, 2, $\cdots$n respectively

and
$$Q_x = k_x i H \tag{c}$$

where k_x = effective coefficient of permeability for flow in the horizontal direction

$$H = H_1 + H_2 + \cdots + H_n$$

Substitution of Eqs. (b) and (c) into Eq. (a) yields

$$k_x H = k_1 H_1 + k_2 H_2 + \cdots + k_n H_n$$

Hence

$$k_x = \frac{1}{H}(k_1 H_1 + k_2 H_2 + \cdots k_n H_n) = \frac{1}{H}\sum_{i=1}^{n} k_i H_i \tag{2.4.4}$$

From Eq. (2.4.4), for flow in the horizontal direction, it can be seen that if the thickness of each soil layer is close whereas the coefficient of permeability is quite different, the value of k_x depends on the most permeable soil layer (k' and H'

are the thickness and permeability coefficient of the most permeable soil layer respectively). Thus the value of k_x is approximate to $k'H'/H$.

(2) Flow in the vertical direction

For flow in the vertical direction for the soil layers shown in Fig. 2.4.4,

$$Q_y = Q_{1y} = Q_{2y} = \cdots = Q_{ny} \tag{a}$$

where $Q_{1y}, Q_{2y}, \cdots, Q_{ny}$ are the discharge velocities in layers 1, 2, $\cdots$n, respectively;

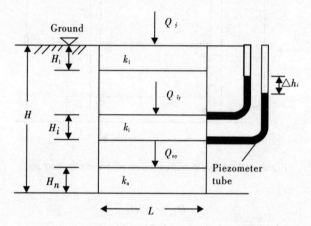

Fig. 2.4.4 Flow in vertical direction in stratified soil

Note that for flow in the vertical direction, the head loss of each soil layer is Δh_i, the hydraulic gradient i_i is $\Delta h_i/H_i$, so

$$Q_{iy} = k_1 i_1 = k_2 i_2 = \cdots = k_n i_n = k_i \frac{\Delta h_i}{H_i} A \tag{b}$$

where $k_1, k_2, \cdots k_n$ = coefficient of permeability for layers 1, 2, $\cdots$n, respectively, for flow in the vertical direction

$i_1, i_2, \cdots i_n$ = hydraulic gradient in soil layers 1, 2, $\cdots$n, respectively

A = the area of cross section of the soil perpendicular to the direction of flow.

The head loss of the entire soil mass h is $\sum \Delta h_i$, the overall average hydraulic gradient i is h/H, so

$$Q_y = k_y \frac{h}{H} A \tag{c}$$

where k_y = effective coefficient of permeability for flow in the vertical direction

Substitution of Eqs. (b) and (c) into Eq. (a) yields

$$k_y = \frac{H}{\sum_{i=1}^{n} \left(\frac{H_i}{k_i}\right)} \tag{2.4.5}$$

From Eq. (2.4.5), for flow in the vertical direction, it can be seen that if the thickness of each soil layer is close whereas the coefficient of permeability is quite different, the value of k_y depends on the most impermeable soil layer (k'' and H'' are the thickness and permeability coefficient of the most impermeable soil layer, respectively). Thus the value of k_x is approximate to $k''H/H''$.

2.4.3 Factors affecting the coefficient of permeability

The coefficient of permeability depends on several factors, most of which are listed as follows.

(1) Shape and size of the soil particles. Permeability increases when the soil particles are much coarser and rounded.

(2) Void ratio. Permeability increases with increase in void ratio. In soil with good gradation, fine soil particles fill in coarse soil particles, making the decrease in void ratio, and eventually the decrease in permeability.

(3) Degree of saturation. Permeability increases with increase in degree of saturation.

(4) Composition of soil particles. For sands and silts, this is not important. However, for soils with clay minerals, this is one of the most important factors. Permeability depends on the thickness of water held to the soil particles, which is a function of the cation exchange capacity, valence of the cations, and so forth. Other factors remaining the same, the coefficient of permeability decreases with increasing thickness of the diffuse double layer.

(5) Soil structure. Fine-grained soils with a flocculated structure have a higher coefficient of permeability than those with a dispersed structure.

(6) Viscosity of the fluid.

(7) Density and concentration of the fluid.

 Flow nets

2.5.1 Definition

A set of flow lines and equipotential lines is called a flow net. A flow line is line along which a water particle will travel. An equipotential line is a line joining the points that show the same piezometric elevation. Figure 2.5.1 shows an example of a flow net for a dam. The permeable layer is isotropic with respect to the coefficient of permeability, i.e., $k_x = k_z = k$. Note that the solid lines in Fig. 2.5.1 are the flow lines and the broken lines are the equipotential lines. In drawing a flow net, the boundary conditions must be kept in mind.

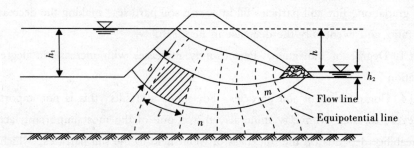

Fig. 2.5.1 A flow net around a dam

It must be remembered that the flow lines intersect with the equipotential lines at right angles. The flow and equipotential lines are usually drawn in such a way that the flow elements are approximately squares.

2.5.2 Calculation of seepage from a flow net under a hydraulic structure

The soil seepage situation can be qualitatively determined by the flow net.

Figure 2.5.1 shows a flow net around a dam. It can be seen that near the draining prism there are the most dense flow lines, showing that the values of hydraulic gradient and seepage velocity are also the largest in this region. Accordingly, the flow lines far away from the seepage prism are quite loose with the relevant smaller values of hydraulic gradient and seepage velocity.

If the quantity of flow between the two consecutive flow lines is the same, the potential energy difference between the two consecutive equipotential lines is also the same, we can get that

(1) The flow line density indicates the different intensities of groundwater flow; the more denser the flow lines are, the greater the flow intensity is.

(2) The equipotential line density indicates the rate of hydraulic gradient change; the more denser the equipotential lines are, the greater the hydraulic gradient is.

Suppose that flow elements are squares, so $b = l$; let dq be the flow through the flow channel per unit length of the hydraulic structure (i.e., perpendicular to the section shown). According to Darcy's law,

$$dq = -b \cdot v = -b \cdot ki = -b \cdot k \frac{dh}{l} \approx -kdh \qquad (2.5.1)$$

Suppose that there are n potential drops, m flow channels in a flow net, the rate of seepage per unit length of the hydraulic structure is

$$q = n \cdot dq = n \cdot kdh = n \cdot k \frac{h}{m} = \frac{n}{m} \cdot kh \qquad (2.5.2)$$

Besides, by use of a flow net, hydraulic head, hydraulic gradient, the seepage quantity, pore water pressure and seepage force at every point in the seepage field can be calculated.

2.6 Seepage force and critical hydraulic gradient

2.6.1 Seepage force

Flow of water through a soil mass results in a certain force being exerted on the soil itself, which is defined as the seepage force, which is a body force and indicated by G_d (kN/m^3). The resistance to the seepage water for soil particles T is also body force, and $T = -G_d$.

To evaluate the seepage force per unit volume a soil, a water column BA is considered, the length and section area of which are L and A, respectively, as shown in Fig. 2.6.1. The weight of the water column is $\gamma_w LA$, the hydrostatic force on the side B is $\gamma_w h_1 A$, the hydrostatic force on the side A is $\gamma_w h_2 A$, the resistance of the seepage water from soil particles is TAL. For equilibrium,

$$\gamma_w h_1 A + \gamma_w LA\cos\alpha - \gamma_w h_2 A - TAL = 0 \tag{a}$$

$$\cos\alpha = \frac{z_1 - z_2}{L} \tag{b}$$

However, $z_1 + h_1 = H_1$, $z_2 + h_2 = H_2$, $i = \dfrac{H_1 - H_2}{L}$, so from Eqs. (a) and (b) yields

$$T = \gamma_w i \tag{c}$$

Hence, the seepage force G_d is

$$G_d = \gamma_w i \tag{2.6.1}$$

The direction of seepage force is the same as the flow direction. From Eq. (2.6.1), G_d is body force.

Sometimes, G_d is expressed by surface force or stress:

$$u = G_d L = \gamma_w iL \tag{2.6.2}$$

where L is the length of flow (m); u is the seepage force or stress (Pa), at the exit of seepage, $u = 0$.

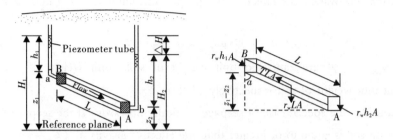

Fig. 2.6.1 Force analysis on the watercolumn

With the change in flow direction, the seepage force will make different effects on the soil, Fig. 2.6.2. When upward seepage occurs, the direction of seepage force and gravity is just opposite, making the effective stress decreases. The soil weight also decreases with the increase in pore water pressure. When downward seepage occurs, the direction of seepage force and gravity is just the same, making the effective stress increases. The soil particles become more compact with the increase in pore water pressure, which is an advantage for engineering construction. For example, in ancient China, there is a method by pouring water in the medium or coarse sand, and applying vibration to make the sand compact. When lateral seepage occurs, it is convenient to calculate pore water pressure by using flow net.

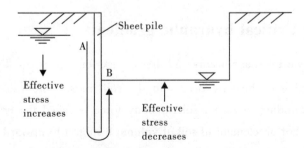

Fig. 2.6.2 Effects of seepage on the effective stress

2.6.2 Upward seepage through soil

Under the water, the effective gravity per unit volume of soil mass is W', that is

$$W' = \gamma_{sat} - \gamma_w = \gamma' \tag{2.6.3}$$

where γ_{sat}, γ_w, γ' are saturated unit weight of soil, unit weight of water and buoyant unit weight of soil respectively (kN/m^3).

Consider the special case of seepage vertically upwards. If the seepage force G_d exerted on soil is equal to or greater than the effective gravity W' ($G_d = \gamma_w \cdot i \geq \gamma'$), the soil loses its strength and behaves like a viscous fluid. The failure caused by seepage occurs. The soil state at which the strength is zero is called static liquefaction. Other name such as heaving, boiling is used to describe specific events connected to the unstable state. Boiling occurs when the upward seepage force exceeds the downward force of the silt. Heaving occurs when seepage forces push the bottom of an excavation upward. If the upward seepage forces exceed the submerged weight, the particles may be carried upwards to be deposited at the ground surface and a "pipe" is formed in the soil near the surface, which is called piping. Piping refers to the subsurface "pipe-shaped erosion". Piping failure can lead to the complete failure of a foundation or to the collapse of earth structure. Thus, it is important to check the potential instability condition under engineering construction.

2.6.3 Critical hydraulic gradient

When upward seepage occurs and the seepage force G_d is equal to the effective gravity W', piping or heaving occurs in the soil mass. The value of hydraulic gradient corresponding to zero resultant body force is called the critical hydraulic gradient (i_{cr}). For an element of soil of volume V subject to upward seepage under the critical hydraulic gradient, the seepage force (Eq. 2.6.1) is therefore equal to the effective weight (Eq. 2.6.3) of the element, i.e.

$$G_d = \gamma_w V \cdot i_{cr} = \gamma' V$$

So

Chapter 2 Permeability of soils and seepage force

$$i_{cr} = \frac{\gamma'}{\gamma_w} \qquad (2.6.4)$$

According to the phase relationships in Chapter 1,

$$\gamma' = (d_s - 1)(1 - n)\gamma_w = \frac{d_s - 1}{1 + e}\gamma_w$$

Therefore,

$$i_{cr} = (d_s - 1)(1 - n) = \frac{d_s - 1}{1 + e} \qquad (2.6.5)$$

The symbols mentioned above have the same indication as in Chapter 1.

The ratio $\frac{\gamma'}{\gamma_w}$ is approximately 1.0 for most soil. When the hydraulic gradient is i_{cr}, the effective normal stress on any plane will be zero, gravitational forces having been cancelled out by upward seepage forces. In the case of sands the contact forces between particles will be zero and the soil will have no strength. It should be realized that "quicksand" is not a special type of soil but simply sand through which there is an upward flow of water under a hydraulic gradient equal to or exceeding i_{cr}.

Example 2.1

As shown in Fig. 2.6.3, the thickness of clay layer under the pit bottom is 5m. There is confined water under the clay layer and the hydraulic pressure height is measured by piezometer tube. During the construction, the groundwater level keeps at the depth 0.5m below the pit bottom by foundation pit dewatering. The unit weight (γ) is 17kN/m³ and the saturated unit weight (γ_{sat}) is 18.6kN/m³ of the clay. Determine whether pit bottom upheaval happens or not.

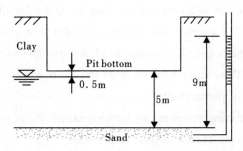

Fig. 2.6.3 Example 2.1

Solution 1

Since confined water exists, there will be upward seepage force, that is

$$G_d = \gamma_w \cdot i = \gamma_w \cdot \frac{\Delta h}{L} = 9.8 \times \frac{4.5}{4.5} = 9.8 \text{kN/m}^3$$

Through 4.5m flow path, the seepage force is expressed by surface force, that is

$$G'_d = G_d \cdot L = 9.8 \times 4.5 = 44.1 \text{kPa}$$

The effective weight of the soil layer above and below the groundwater is

$$W' = 17 \times 0.5 + (18.6 - 9.8) \times 4.5 = 48.1 \text{kPa}$$

Since $G_d \leqslant W'$, pit bottom upheaval can't happen.

Solution 2

The Pressure head difference of the confined water is $H = 9 - 4.5 = 4.5$m. Hence,

$$\gamma_w H = 9.8 \times 4.5 = 44.1 \text{kPa}$$

The effective weight of the soil layer between the pit bottom and the top surface of confined water is

$$\sum \gamma_i h_i = 17 \times 0.5 + (18.6 - 9.8) \times 4.5 = 48.1 kPa$$

Since $\sum \gamma_i h_i > \gamma_w H$, pit bottom upheaval can't happen.

EXERCISES

2.1 What are the main influences of capillary water on buildings and soil?

2.2 What is Darcy's Law and how to use it for different kinds of soil?

2.3 What is the principle of constant-head test to determine the coefficient of permeability? Why should falling-head test be used in clay soil to determine the coefficient of permeability?

2.4 What is the forming condition of heaving sand? What are the main damages caused by heaving sand and how to prevent?

2.5 In a simple constant-head permeability testing apparatus (Fig. 2.1), the sectional area of the specimen is 120cm^2. The water through the specimen is measured by the measuring cylinder. After 10s the amount of water is 150cm³. Determine the coefficient of permeability of the specimen.

2.6 A certain cohesiveless soil has the void ratio $e = 0.61$ and the specific gravity $d_s = 2.65$. When the heaving sand just happens, determine the critical hydraulic gradient.

2.7 In the excavation of a certain foundation pit, draining water causes upward seepage. The water head difference is 60cm, the length of water flow through the soil is 50cm, the saturated unit weight $\gamma_{sat} = 20.5$ kN/m³. Determine whether the heaving sand happens or not.

2.8 One clay layer is located between two sand layers. The bulk unit weight of the sand (γ) is 17.6 kN/m³, the saturated unit weight of the sand is (γ_{sat}) 19.6 kN/m³, the saturated unit weight of the clay (γ_{sat}) is 20.6 kN/m³. The thickness of each soil layer is shown in Fig. 2.2. The groundwater level is at the depth 1.5m below the ground level. Assuming that there is pressured water in the lower sand layer, and the water level of the standpipe is at 3m above the ground level, then if heaving occurs at the clay layer, How much is the water level of the standpipe higher than the ground level?

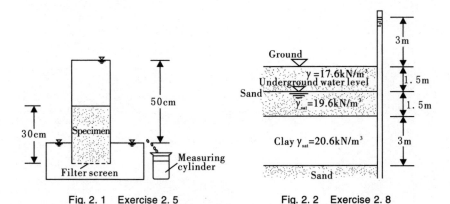

Fig. 2.1 Exercise 2.5   Fig. 2.2 Exercise 2.8

REFERENCES

1. Karl Terzaghi (1943). Theoretical Soil Mechanics[M]. John Wiley and Sons, New York.
2. T. William Lambe and Robert V. Whitman (1969). Soil Mechanics[M]. John Wiley and Sons, New York.
3. John Atkinson (1993). An Introduction to the Mechanics of Soils and Foundations[M]. McGRAW-HILL Book Company Europe, England.
4. R. F. Craig (1998). Soil Mechanics[M]. E and FN Spon, London and New York.
5. Soil mechanics work team at Hohai University (2004). Soil Mechanics[M]. China Communication Press, Beijing.

SOIL MECHANICS
CHAPTER 3
Stress distribution in soils

3.1 Introduction

◀ 3.1.1 Purpose of researching the stress state of soil

Geotechnical engineering is the application of the science of soil mechanics and rock mechanics, engineering geology and other related disciplines to the civil engineering construction, the extractive industries (e. g., underground mining) and the preservation and enhancement of the environment (e. g., landfills of municipal solid waste or nuclear waste) (Anon, 1999).

Geotechnical engineering is a subdiscipline within civil engineering. It covers all forms of soil-related problems. Geotechnical engineering plays a key role in all civil engineering projects, since all construction is built on or in the soil or rock on earth or on planets (e. g., the unique experiences involved in trying to obtain remotely the properties of the granular material on the moon and Mars were summarized by Scott, 1987). The soil or rock formation supporting every artificial structure is called the ground. The ground is commonly divided into two broad categories. One is the natural ground that consists of intact natural soils or rocks. The other is the man-made ground that consists of artificially improved soils or rocks. The foundation of a structure is defined as that part of the structure in direct contact with the ground and which transmits the load of the structure to the ground without overstressing the ground (Tomlinson, 1994). It is obvious that when the load of the structure is applied to the ground via the foundation, stresses and deformations will be induced in the ground. This would arise two engineering problems related to the structure, i. e., the soil stability problem and the soil deformation problem. On the one hand, if the shear stresses induced in the ground are within the allowable rang of the soil strength, the soil is stable. Conversely if the shear stresses induced in a certain localized area in the ground are in excess of the allowable range of the soil strength, failure of the soil will take place there, and this may result in the slippage of the global ground, leading to an overturning of the structure. On the other hand, if the deformations of the ground exceeds allowable values, the structure may also be damaged and lose its service function, although the soil has not failed yet. Therefore, in order to ensure the safety and the normal service function of the structure, the stress distribution patterns and the deformations induced in the ground under various loading conditions must be studied.

Calculation of the ground deformations is introduced in the next chapter. This chapter presents only the calculation of the stresses in the ground soils and their distribution patterns. The main content of this chapter is based on a Chinese textbook written by Zhao (2010), and it is supplemented with two English textbooks written by Whitlow (2001) and Craig (1997) respectively.

Actually, the stresses in the ground soils can be divided into two categories in terms of their origins:

One is the effective overburden pressure, i. e., the stresses due to the effective self-weight of the soil. It is also known as the geostatic pressure or the at-rest in situ stress. Generally speaking, compression of the soils caused by their effective self-weights has been already completed over a very long period of geological history, such that soil deformation due to the effective overburden pressure would not be induced further. However, it is an exception for newly deposited, sedimentation or reclaimed soils such as recent hydraulic fills.

The other is the stress increase, i. e., the stress in the interior of the ground due to exterior (static or dynamic) load. It is also known as the additional stress. The stress increase may be the major cause of soil instability and ground deformations. The magnitude of the stress increase is dependent not only on the location of the calculation point, but also on the magnitude and distribution pattern of the contact pressure between the foundation and the ground.

This chapter presents firstly the calculation of the effective overburden pressure, followed then by the distribution pattern and the corresponding calculation method of the contact pressure beneath the foundation. Finally introduced is the calculation of the stress increase due to various loading conditions.

3.1.2 Stress state of soil

1. The basic assumption of the stress calculation

(1) Continuous medium

Elastic theory request that soil is continuous medium. In fact soil is comprised of three phase matter.

(2) Linear elastic body

When the stress is small, the stress-strain relationship of soil is linear.

(3) Homogeneous and isotropic body

When variation of the property of soil layer is not too large, soil is a homogeneous and isotropic body.

We can adopt the solutions in elastic mechanics to solve the stress in elastic soil and the analysis method could be simple and easy to draw graph.

2. The common stress state in ground

(1) The general state of stress—three-dimensional problem

The stress state of groundwork is three-dimensional problem under load acting, an elemental soil mass with sides measuring dx, dy, and dz is shown in Fig. 3.1.1. Parameters σ_x, σ_y, and σ_z are the normal stress acting on the planes normal to the x, y, and z axes, respectively. The normal stresses are considered positive when they are directed onto the surface. Parameters τ_{xy}, τ_{yx}, τ_{yz}, τ_{zy}, τ_{xz}, and τ_{zx} are shear stresses. All shear stresses are positive in Fig. 3.1.1.

The stress at any point is a function of and the stress state at the point has nine stress components, i.e., and write a matrix form as follows

$$\sigma_{ij} = \begin{bmatrix} \sigma_x & \tau_{xy} & \tau_{xz} \\ \tau_{yx} & \sigma_y & \tau_{yz} \\ \tau_{zx} & \tau_{zy} & \sigma_z \end{bmatrix} \quad (3.1.1)$$

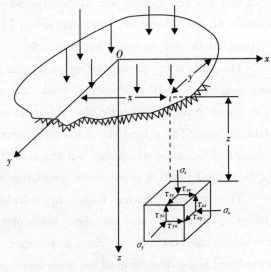

Fig. 3.1.1 The general stress state of soil

For equilibrium, $\tau_{xy} = \tau_{yx}$, $\tau_{yz} = \tau_{zy}$, $\tau_{xz} = \tau_{zx}$, the elemental soil mass has six stress components, i.e., σ_x, σ_y, σ_z, τ_{xy}, τ_{xz}, and τ_{yz}.

(2) Plane strain conditions—two dimensional problem

There is enough length along the extent direction, the ratio of the length l to the width b of the foundation, namely, $l/b \geq 10$. The geometry of the random fracture which is cut against to y-axis is same and also the stress state in the ground is same and this is the condition of plane strain problem. In the plane strain conditions, soil has deformation in the plane of x and z, but not in the direction of y, i.e., $\varepsilon_y = 0$ and $\tau_{yx} = \tau_{yz} = 0$ because of symmetrical characteristics. The unit body has five stress component, i.e., $\sigma_x, \sigma_y, \sigma_z, \tau_{xz}, \tau_{zx}$, and write a matrix form as follows:

$$\sigma_{ij} = \begin{bmatrix} \sigma_x & 0 & \tau_{xz} \\ 0 & \sigma_y & 0 \\ \tau_{zx} & 0 & \sigma_z \end{bmatrix} \qquad (3.1.2)$$

(3) State of the confining stress—one-dimensional problem

The horizontal ground is a semi-infinite half-space body and the geostatic stress in semi-infinite elastic ground is only relate to z, namely, soil point unit or soil element hasn't lateral motion and this is defined as the state of the confining stress. Thus, the random vertical plane is symmetrical plane, so $\tau_{xy} = \tau_{yz} = \tau_{zx} = 0$, the stress matrix is given by

$$\sigma_{ij} = \begin{bmatrix} \sigma_x & 0 & 0 \\ 0 & \sigma_y & 0 \\ 0 & 0 & \sigma_z \end{bmatrix} \qquad (3.1.3)$$

where $\sigma_x = \sigma_y$ due to $\varepsilon_x = \varepsilon_y = 0$, and it is a direct ratio with respect to σ_z.

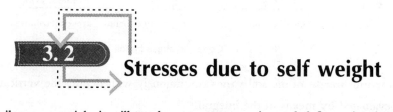

3.2 Stresses due to self weight

Soil as a material, it will produce stress at a point and deformation in the soil when it is subjected to loading. Generally, data concerning internal stress conditions are used to determine deformation, i.e. the deformation can be determined by using stress. Stresses within soil are caused by the external loads applied to the soil and by the weight of the soil. The pattern of stresses caused by applied loads is usually quite complicated.

(1) Vertical geostatic stress

In the situation just described, there are no shear stresses upon vertical and horizontal planes within the soil. Hence the vertical geostatic stress at any depth can be computed simply by considering the weight of soil above that depth.

Thus, if the unit weight of the soil is constant with depth

$$\sigma_z = \gamma_1 h \qquad (3.2.1)$$

where h is the depth and γ is the total unit weight of the soil. In this case, the vertical stress will vary linearly with depth, as shown in Fig. 3.2.1(a). The distribution of vertical stress along the depth is a triangle. If the soil is stratified and the unit weight is different for each stratum, then the vertical stress can conveniently be computed by means of the summation, as shown in Fig. 3.2.1(b).

$$\sigma_z = \sum_{i=1}^{n} \gamma_i h_i \qquad (3.2.2)$$

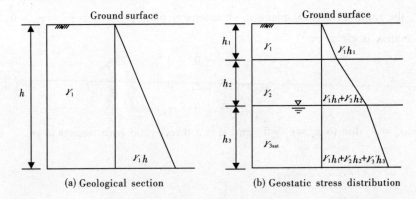

Fig. 3.2.1 Geostatic stress in soil

If the unit weight of the soil varies continuously with depth, the vertical stress can be evaluated by means of the integral.

Example 3.1

A foundation is composed by multi-layer soils, which is shown in the geological section, Fig. 3.2.2(a). Calculate and draw geostatic stress distribution of the soils along the depth direction.

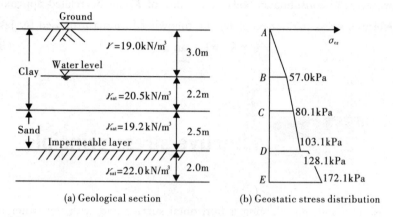

(a) Geological section   (b) Geostatic stress distribution

Fig. 3.2.2 Example 3.1

Solution

Take five points A, B, C, D, E as the reference points to calculate the geostatic stress, shown as Fig. 3.2.2(b).

A: $\sigma_{czA} = 0$

B: $\sigma_{czB} = \gamma_B h_B = 19 \times 3 = 57.0 \text{kPa}$

C: $\sigma_{czC} = \gamma_B h_B + \gamma'_C h_C = 19 \times 3 + (20.5 - 10) \times 2.2 = 80.1 \text{kPa}$

D up: $\sigma_{czDup} = \gamma_B h_B + \gamma'_C h_C + \gamma'_{Dup} h_D = 80.1 + (19.2 - 10) \times 2.5 = 103.1 \text{kPa}$

D down: $\sigma_{czDdown} = \gamma_B h_B + \gamma'_C h_C + \gamma_{satDdown} h_D = 80.1 + 19.2 \times 2.5 = 128.1 \text{kPa}$

E: $\sigma_{czE} = \gamma_B h_B + \gamma'_C h_C + \gamma_{satDdown} h_D + \gamma_{satE} h_E = 128.1 + 22 \times 2 = 172.1 \text{kPa}$

(2) Horizontal geostatic stress

The ratio of horizontal to vertical stress for a mass of soil in a state of rest is expressed by a factor called the coefficient of earth pressure at rest or lateral stress ratio at rest, and is denoted by the symbol K_0, i.e.

$$K_0 = \frac{\sigma_{h_0}}{\sigma_{v_0}} \qquad (3.2.3)$$

The value of K_0 can be determined experimentally by means of a triaxial test in which the axial stress and the all-round pressure are increased simultaneously such that the lateral strain in the specimen is maintained at zero (the hydraulic triaxial apparatus is most suitable for this purpose).

For normally consolidated soils, the value of K_0 can be related approximately to the effective stress parameter φ' by the following formula proposed by Jaky.
$$K_0 = 1 - \sin \varphi' \tag{3.2.4}$$

3.3 Effective stress principle

Consider a soil mass having a horizontal surface and with the water table at surface level. The total vertical stress (i.e. the total normal stress on a horizontal plane) at depth h is equal to the weight of all material (solids + water) per unit area above that depth.

A soil can be visualized as a skeleton of solid particles enclosing continuous voids which contain water and/or air. The volume of the soil skeleton as a whole can change due to rearrangement of the soil particles into new positions, mainly by rolling and sliding, with a corresponding change in the forces acting between particles. The actual compressibility of the soil skeleton will depend on the structural arrangement of the solid particles. In a fully saturated soil, since water is considered to be incompressible, a reduction in volume is possible only if some of the water can escape from the voids. In a dry or a partially saturated soil a reduction in volume is always possible due to compression of the air in the voids, provided there is scope for particle rearrangement.

Shear stress can be resisted only by the skeleton of solid particles, by means of forces developed at the interparticle contacts. Normal stress may be resisted by the soil skeleton through an increase in the interparticle forces. If the soil is fully saturated, the water filling the voids can also withstand normal stress by an increase in pressure.

Chapter 3 Stress distribution in soils

The importance of the forces transmitted through the soil skeleton from particle to particle was recognized in 1923 when Terzaghi presented the principle of effective stress, an intuitive relationship based on experimental data. The principle applies only to fully saturated soils and relates the following three stresses.

(1) The effective normal stress (σ') on the plane, representing the stress transmitted through the soil skeleton only. It is the total particle interaction force per unit area. It controls: (a) deformation and (b) shear strength of soil mass.

(2) The total normal stress (σ) on a plane within the soil mass, being the total (particle + water) force per unit area.

(3) The pore water pressure (u), being the pressure of the water filling the void space between the solid particles;

The relationship is

$$\sigma' = \sigma - u \tag{3.3.1}$$

The principle can be represented by the following physical model. Consider a 'plane a-a' in a fully saturated soil, passing through points of interparticle contact only, as shown in Fig. 3.3.1. Then, the effective normal stress is interpreted as the sum of all the components N' within the area A, divided by the area A, i.e.

$$\sigma' = \frac{\sum N'}{A} \tag{3.3.2}$$

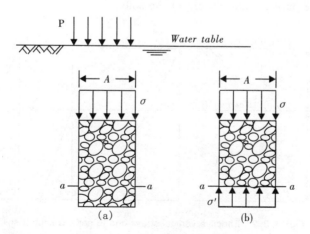

Fig. 3.3.1 Interpretation of effective stress

The total normal stress is given by

$$\sigma = \frac{P}{A} \quad (3.3.3)$$

If point contact is assumed between the particles, the pore water pressure will act on the plane over the entire area A. Then, for equilibrium in the direction normal to a-a

$$P = \sum N' + uA \quad \text{or} \quad \frac{P}{A} = \frac{\sum N'}{A} + u \quad \text{i.e.} \quad \sigma = \sigma' + u \quad (3.3.4)$$

Consider a soil mass having a horizontal surface and with the water table at surface level, as shown in Figure 3.3.2. The total vertical stress (i.e. the total normal stress on a horizontal plane) at depth C is equal to the weight of all material (solids + water) per unit area above that depth, i. e.

$$\sigma_v = \gamma h_1 + \gamma_{sat} h_2 \quad (3.3.5)$$

The pore water pressure at any depth will be hydrostatic since the void space between the solid particles is continuous, so at depth C

$$u = \gamma_w h_2 \quad (3.3.6)$$

Hence, from the Terzaghi's Effective Stress Principle the effective vertical stress at depth C will be

$$\sigma'_v = \sigma_v - u = \gamma h_1 + (\gamma_{sat} - \gamma_w) h_2 = \gamma h_1 + \gamma' h_2 \quad (3.3.7)$$

where γ' is the buoyant unit weight of the soil.

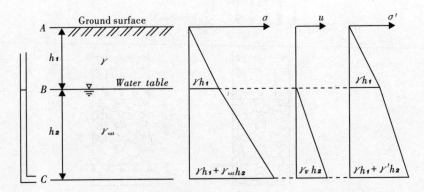

Fig. 3.3.2 Effective vertical stress due to self-weight of soil

Example 3.2

A foundation is composed by multi-layer soils, which is shown in the geological section, Fig. 3.3.3 (a). Calculate and draw total normal stress distribution, pore water pressure distribution and effective normal stress distribution of the soils along the depth direction.

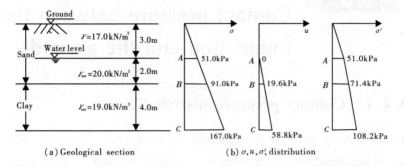

(a) Geological section

(b) σ, u, σ' distribution

Fig. 3.3.3 Example 3.2

Solution

Take three points A、B、C as the reference points to calculate σ, u, σ', shown in Table 3.3.1.

Table 3.3.1 Calculation of example 3.2

Reference Points	σ (kN/m²)	u (kN/m²)	σ' (kN/m²)
A	3 ×17 =51.0	0	51
B	(3 ×17) +(2 ×20) =91.0	2 ×9.8 =19.6	71.4
C	(3 ×17) +(2 ×20) +(4 ×19) =167.0	6 ×9.8 =58.8	108.2

3.4 Contact pressure between the foundation and the ground

3.4.1 Contact pressure distribution

Virtually every structure is supported by ground soils or rocks. The major function of the foundation of structure is to transmit the load of the structure to the supporting ground. Therefore, the analysis of the interaction between a structural foundation and the supporting ground soil media is of primary importance to both structural and geotechnical engineering.

Contact pressure is the intensity of loading transmitted from the underside of a foundation to the ground soil (Whitlow, 2001). The magnitude and the distribution pattern of the contact pressure have an important impact on the stress increase induced in the ground. The magnitude and the distribution pattern of the contact pressure depend on many factors such as the magnitude and distribution of the structure load applied, the rigidity and embankment depth of the foundation, and the soil properties, etc.

It has been found from tests that for a foundation with a very low rigidity or for a flexible foundation, the magnitude and the distribution pattern of the contact pressure is the same as those of the load applied on the foundation. This is because the foundation is compatible to the deformation of the ground soil. When the load on the foundation is uniformly distributed, the contact pressure (normally denoted as the reaction force on the underside of a foundation, ditto) is also uniformly distributed, as shown in Fig. 3.4.1(a). When the load distribution is trapezoidal, the contact pressure distribution is also trapezoidal, as shown in Fig. 3.4.1(b).

For a rigid foundation that cannot be compatible to the ground deformation due to significant difference in rigidity, the distribution of the contact pressure varies with the magnitude of the applied load, the embedment depth of the

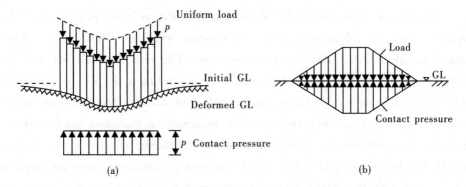

Fig. 3.4.1 Contact pressure distribution beneath a flexible foundation

foundation, and the properties of the ground soil. For instance, when a centric load is applied to the rigid strip foundation founded on the surface of a sandy ground, the contact pressure on the centerline of the foundation is the maximum, the contact pressure on the edge of the foundation is zero, and its distribution looks like a parabolic curve, as shown in Fig. 3.4.2(a). This is because no cohesion is available among sand particles. When a centric load is applied to the rigid strip foundation founded on the surface of a clayey ground, some loads can be carried on the edge of the foundation due to cohesion of the clayey soils. Therefore, when the applied load is relatively small, high contact pressure will be imposed on the edge of the foundation and low contact pressure on the center of the foundation. The distribution curve looks like a saddle-shape. When the load is increased gradually to failure load, the distribution curve of the contact pressure becomes higher on the center and lower on the edge of the foundation, like a bowl—shape, as shown in Fig. 3.4.2(b).

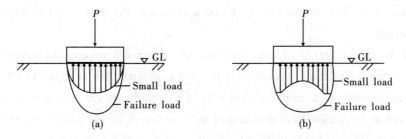

Fig. 3.4.2 Contact pressure distribution beneath a rigid foundation

Empirically, when the width of the rigid foundation is not too wide and the applied load is relatively small, the contact pressure distribution follows approximately a linear distribution assumption. The error induced between the assumption and the reality would be acceptable, according to St. Venant's principle (If forces acting on a small portion of the surface of an elastic body are replaced by another statically equivalent system of forces acting on the same portion of the surface, this redistribution of loading produces substantial changes in the stresses locally but has a negligible effect on the stresses at distances which are large in comparison with the linear dimensions of the surface on which the forces are changed). (Timoshenko and Goodier, 1951). Introduced below is the simplified calculation method normally used in the engineering practice for computing the contact pressure based on the linear distribution assumption.

◀ 3.4.2 Contact pressure due to vertical centric load

The length and width of a rectangular foundation are l and b, respectively, as shown in Figs. 3.4.3a and 3.4.3b. A vertical centric load F is applied on the foundation. According to the linear distribution assumption, the value of the contact pressure is

$$p = \frac{F+G}{A} = \frac{F+G}{l \times b} \quad (3.4.1)$$

where lowercase p represents the contact pressure (kPa); F represents the vertical load on the upside of the foundation (kN); G represents the self-weight of the foundation and the soil weight on the steps of foundation, generally, the value 20kN/m^3 is adopted as the average unit weight of them; and $A = l \times b$ represents the area of the foundation (m^2). l and b represent the length and width of the foundation.

If the foundation is oblong (theoretically when l/b approaches infinity it is called a strip foundation, practically when l/b is greater than or equal to 10 it can be taken granted as a strip foundation), a free-body of 1m unit length can be truncated in the longitudinal direction of the foundation for the calculation analysis, as shown in Fig. 3.4.3(c). In this situation, the contact pressure is

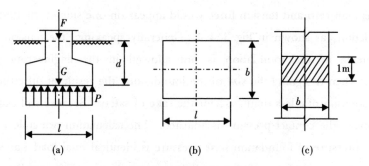

Fig. 3.4.3 Contact pressure distribution due to vertical centric load

$$p = \frac{F+G}{b} \quad (3.4.2)$$

where b represents the width of the foundation (m); the other symbols have the same meanings as presented before.

3.4.3 Contact pressure due to one-way vertical eccentric load

When an one-way eccentric load is applied to a rectangular foundation (as shown in Fig. 3.4.4), the contact pressure at any arbitrary point can be calculated using the formula of eccentric compression in mechanics of materials, as given by

$$p_{\substack{max \\ min}} = \frac{F+G}{A} \pm \frac{M}{W} = \frac{F+G}{A}(1 \pm \frac{6e}{l}) \quad (3.4.3)$$

where p_{max} and p_{min} represent the maximum and minimum contact pressures on both sides of the underside of the foundation (kPa); M represents the moment of the eccentric load about Y-Y axis; W represents resisting moment of the underside of foundation, $W = \frac{bl^2}{6}$ if the area is rectangular (m³); e is the offsetting of the eccentric load line to the Y-Y axis.

It can be seen from Eq. (3.4.3) that when the resultant offsetting e is less than $l/6$, the distribution curve of the contact pressure is trapezoidal. When the resultant offsetting e is equal to $l/6$, p_{min} is zero and the distribution curve of the contact pressure is triangular. When the resultant offsetting e is greater than $l/6$,

p_{min} is less than zero and tension force would appear on one side of the underside of the foundation, as shown in Fig. 3.4.4. Generally speaking, the tension force on the underside of the foundation is not allowed in the engineering practice; therefore, when designing the size of the foundation, the resultant offsetting should satisfy a criterion of e less than $l/6$, for the sake of safety. Because soil cannot bear tension force, the contact pressure is adjusted. The calculation principle is that the composite pressure of foundation base pressure is identical total load (as shown in Fig. 3.4.4d), the formula of the maximum contact pressure p_{max} is given by

$$p_{max} = \frac{2(F+G)}{3ba} \qquad (3.4.4)$$

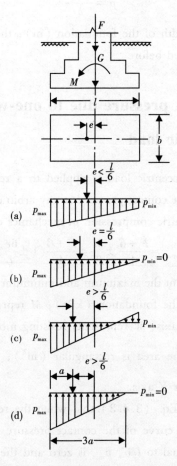

Fig. 3.4.4 Contact pressure distribution due to one-way vertical eccentric load

where a represents the distance between the action point of eccentric load and the edge of the p_{max}, $a = \dfrac{l}{2} - e$, (m)

Similarly for a strip foundation, the maximum and minimum contact pressures of the underside of the foundation are

$$p_{\substack{max\\min}} = \dfrac{F+G}{A}\left(l \pm \dfrac{6e}{b}\right) \qquad (3.4.5)$$

◀ 3.4.4 Contact pressure due to two-way vertical eccentric load

When a two-way eccentric load is applied to a rectangular foundation (as shown in Fig. 3.4.5), the contact pressure at any arbitrary point can be calculated using the formula of eccentric compression in mechanics of materials, as given by

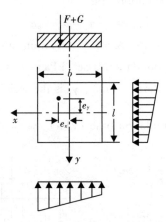

Fig. 3.4.5 Contact pressure distribution due to two-way vertical eccentric load

$$p_{\substack{max\\min}} = \dfrac{F+G}{A} \pm \dfrac{M_x y}{I_x} \pm \dfrac{M_y x}{I_y} \qquad (3.4.6)$$

where $M_x = (F+G)e_y$ represents the moment of the eccentric load about X-X axis (e_y is the offsetting of the eccentric load line to X-X axis); $M_y = (F+G)e_x$ represents the moment of the eccentric load about Y-Y axis (e_x is the offsetting of the eccentric load line to Y-Y axis); $I_x = bl^3/12$ represents the moment of inertia of the area of the underside of the foundation about the X-X axis; $I_y = lb^3/12$ represents the moment of inertia of the area of the underside of the foundation about the Y-Y axis.

3.4.5 Additional stress of the foundation underside

The additional stress is defined as the increased pressure in the foundation due to building the architecture, as shown in Fig. 3.4.6.

1. When the foundation is constructed above the ground surface (Fig. 3.4.6a), the additional stress of the foundation underside p_0 is the contact pressure of the foundation underside p, that is

$$p_0 = p \quad (3.4.7)$$

2. When the foundation is constructed at some depth under the ground surface (Fig. 3.4.6b), the additional stress of the foundation underside p_0 is calculated by the following equation, i.e.

$$p_0 = p - \sigma_c = p - \gamma_0 d \quad (3.4.8)$$

where p is the contact pressure of the foundation underside (kPa); σ_c is the overburden pressure at the foundation base (kPa); d is the depth from the ground surface to the foundation underside (m); γ_0 is the weighted average unit weight of the soil layers above the foundation base (kPa), $\gamma_0 = \dfrac{\sum \gamma_i h_i}{d}$.

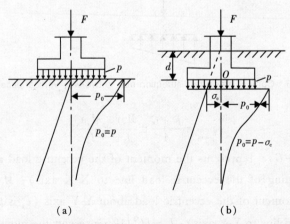

Fig. 3.4.6 Calculation diagram of the additional stress of the foundation underside

Chapter 3　Stress distribution in soils

The increased pressure caused by building is added after earth excavation, so it is the contact pressure of the foundation underside subtracts the original overburden pressure.

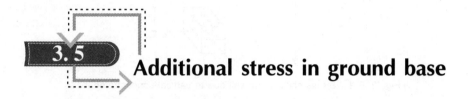

3.5　Additional stress in ground base

Currently in order to obtain the stress increase in the ground due to additional load, the ground soil is generally assumed to be a continuous, homogeneous, isotropic and fully elastic material. The stress increase can then be calculated using the basic formulate of the elasticity theory. In addition, the stress increase problem can be divided, in terms of their natures, into spatial (three-dimensional) problems and plane (two-dimensional) problems. If the stress is a function of the three coordinates x, y, and z, the stress increase problem is a spatial problem. The calculation of the stress increase on the undersides of rectangular and circular foundations etc. is a spatial problem. If the stress is a function of the two coordinates x and y, the stress increase problem is a plane problem. The calculation of the stress increase on the underside of a strip foundation belongs to this category. The foundations of most of the water reservoir engineering constructions such as dams and retaining walls etc. are also strip foundation.

3.5.1　Stress increase due to vertical concentrated (point) load

When a vertical concentrated load F is applied on the surface of an elastic half-space as shown in Fig. 3.5.1, the six stress components $\sigma_x, \sigma_y, \sigma_z, \tau_{xy}=\tau_{yx}$, $\tau_{yz}=\tau_{zy}, \tau_{xz}=\tau_{zx}$ at an arbitrary point M in the interior of the elastic body can be using the elasticity theory, as given by

$$\sigma_z = \frac{3F}{2\pi} \cdot \frac{z^3}{R^5} \qquad (3.5.1a)$$

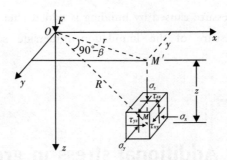

Fig. 3.5.1 Stress state of the soil due to vertical concentrated load

$$\sigma_y = \frac{3F}{2\pi} \cdot \left\{ \frac{y^2 z}{R^5} + \frac{1-2v}{3} \left[\frac{1}{R(R+z)} - \frac{(2R+z)y^2}{(R+z)^2 R^3} - \frac{z}{R^3} \right] \right\} \quad (3.5.1\text{b})$$

$$\sigma_x = \frac{3F}{2\pi} \cdot \left\{ \frac{x^2 z}{R^5} + \frac{1-2v}{3} \left[\frac{1}{R(R+z)} - \frac{(2R+z)x^2}{(R+z)^2 R^3} - \frac{z}{R^3} \right] \right\} \quad (3.5.1\text{c})$$

$$\tau_{xy} = \frac{3F}{2\pi} \cdot \left[\frac{xyz}{R^5} + \frac{1-2v}{3} \cdot \frac{(2R+z)xy}{(R+z)^2 R^3} \right] \quad (3.5.1\text{d})$$

$$\tau_{zy} = \frac{3F}{2\pi} \cdot \frac{yz^2}{R^5} \quad (3.5.1\text{e})$$

$$\tau_{zx} = \frac{3F}{2\pi} \cdot \frac{xz^2}{R^5} \quad (3.5.1\text{f})$$

The above equations are the well-known Boussinesq solution. They are basic formulae for solving the stress increase in the ground. In soil mechanics (as against in elasticity), a uniform sign convention has been used in that the following are considered as positive: compressive stress, reduction in length or volume, and displacement in the positive coordinate direction.

In soil mechanics, the vertical (or normal) stress component σ_z on the horizontal plane is of special importance, as it is the major cause of the compressive deformation of the ground soil. Therefore, discussed below are primarily the calculation of the additional stress and the analyses of its distribution pattern.

In the light of the geometrical relationship $R^2 = r^2 + z^2$ in Fig. 3.5.1, the Eq. (3.5.1a) can be rewritten as given in the following form,

Chapter 3 Stress distribution in soils

$$\sigma_z = \frac{3F}{2\pi} \cdot \frac{z^3}{R^5} = \frac{3F}{2\pi \cdot z^2} \cdot \frac{1}{\left[1+\left(\frac{r}{z}\right)^2\right]^{\frac{5}{2}}} = \alpha \cdot \frac{F}{z^2} \qquad (3.5.2)$$

where $\alpha = \frac{3}{2\pi} \cdot \frac{1}{\left[1+\left(\frac{r}{z}\right)^2\right]^{\frac{5}{2}}}$ is the coefficient of the additional stress under the foundation underside due to vertical concentrated load. It is a function of r/z and it can be read off in Table 3.5.1.

Table 3.5.1 The coefficient of the additional stress α due to vertical concentrated load

r/z	α	r/z	α	r/z	α	r/z	α	r/z	α	r/z	α	r/z	α
0.00	0.4775	0.30	0.3849	0.60	0.2214	0.90	0.1083	1.20	0.0513	1.50	0.0251	2.00	0.0085
0.02	0.4770	0.32	0.3742	0.62	0.2117	0.92	0.1031	1.22	0.0489	1.54	0.0225	2.10	0.0070
0.04	0.4756	0.34	0.3632	0.64	0.2024	0.94	0.0981	1.24	0.0466	1.58	0.0209	2.20	0.0058
0.06	0.4732	0.36	0.3521	0.66	0.1934	0.96	0.0933	1.26	0.0443	1.60	0.0200	2.40	0.0040
0.08	0.4699	0.38	0.3408	0.68	0.1846	0.98	0.0887	1.28	0.0422	1.64	0.0183	2.60	0.0029
0.10	0.4657	0.40	0.3294	0.70	0.1762	1.00	0.0844	1.30	0.0402	1.68	0.0167	2.80	0.0021
0.12	0.4607	0.42	0.3181	0.72	0.1681	1.02	0.0803	1.32	0.0384	1.70	0.0160	3.00	0.0015
0.14	0.4548	0.44	0.3068	0.74	0.1603	1.04	0.0764	1.34	0.0365	1.74	0.0147	3.50	0.0007
0.16	0.4482	0.46	0.2955	0.76	0.1527	1.06	0.0727	1.36	0.0348	1.78	0.0135	4.00	0.0004
0.18	0.4409	0.48	0.2843	0.78	0.1455	1.08	0.0691	1.38	0.0332	1.80	0.0129	4.50	0.0002
0.20	0.4329	0.50	0.2733	0.80	0.1386	1.10	0.0658	1.40	0.0317	1.84	0.0119	5.00	0.0001
0.22	0.4242	0.52	0.2625	0.82	0.1320	1.12	0.0628	1.42	0.0302	1.88	0.0109		
0.24	0.4151	0.54	0.2518	0.84	0.1257	1.14	0.0595	1.44	0.0288	1.90	0.0106		
0.26	0.4054	0.56	0.2414	0.86	0.1196	1.16	0.0567	1.46	0.0275	1.94	0.0097		
0.28	0.3954	0.58	0.2313	0.88	0.1138	1.18	0.0539	1.48	0.0263	1.98	0.0089		

It can be seen from Eq. (3.5.2) that the following three conclusions can be obtained, as follows:

(1) On the concentrated load line ($r=0, \alpha = \frac{3}{2\pi}, \sigma_z = \frac{3}{2\pi} \cdot \frac{P}{z^2}$), the additional stress decreases with increasing depth z, as shown in Fig. 3.5.2.

(2) At a certain distance r away from the concentrated load line, the additional stress σ_z is zero at the ground surface, and it increases gradually with

increasing depth. However, σ_z decreases with increasing depth, as shown in Fig. 3.5.2.

(3) On a horizontal plane at a certain depth z, the additional stress decreases with increasing r, as shown in Fig. 3.5.2.

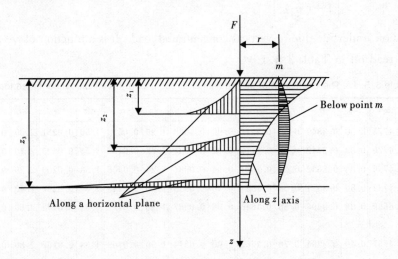

Fig. 3.5.2 Distribution of stress increase due to point load

◀ 3.5.2 Additional stress under the corners of a rectangular foundation underside due to a vertical uniform load

When a vertical uniform load (hereby referred to as the compressive stress, ditto) is applied to the underside of a rectangular foundation, the additional stress under the corners of the foundation can be calculated by integrating the basic Eq. (3.5.2) with respect to the whole rectangular area, as shown in Fig. 3.5.3. If the vertical uniform load intensity on the foundation underside is p, the acting force dp on the infinitesimal area dxdy is pdxdy and it can be taken granted as a concentrated point load. Therefore, the additional stress at a depth z under the foundation corner O induced by this point load is given by

Chapter 3 Stress distribution in soils

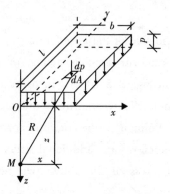

Fig. 3.5.3 Rectangular foundation underside objected to vertical uniform load

$$d\sigma_z = \frac{3p}{2\pi} \cdot \frac{1}{\left[1+\left(\frac{r}{z}\right)^2\right]^{5/2}} \cdot \frac{dxdy}{z^2} \qquad (3.5.3)$$

Substituting $r^2 = x^2 + y^2$ into the above equation and integrating it with respect to the whole area of the foundation underside given the additional stress at depth z under the corner O of the underside of the rectangular foundation induced by the vertical uniform load, as expressed by

$$\sigma_z = \int_0^b \int_0^l \frac{3p}{2\pi} \cdot \frac{z^3 dxdy}{\left(\sqrt{x^2+y^2+z^2}\right)^5}$$

$$= \frac{p}{2\pi} \left[\frac{mn}{\sqrt{1+m^2+n^2}} \cdot \left(\frac{1}{m^2+n^2} + \frac{1}{1+n^2}\right) + arctan\left(\frac{m}{n\sqrt{1+m^2+n^2}}\right) \right]$$

$$= \alpha_c p \qquad (3.5.4)$$

where α_c is the coefficient of the additional stress under the corner O of the underside of a rectangular foundation due to vertical uniform load. It is a function of $m\ (=l/b)$ and $n\ (=z/b)$, namely,

$$\alpha_c = f(m,n) = \frac{p}{2\pi} \left[\frac{mn}{\sqrt{1+m^2+n^2}} \cdot \left(\frac{1}{m^2+n^2} + \frac{1}{1+n^2}\right) + arctan\left(\frac{m}{n\sqrt{1+m^2+n^2}}\right) \right]$$

and it can be read off in Table 3.5.2, where l is the length of the longer side of the foundation underside and b is the width of the shorter side of the foundation underside.

For any point inside and outside the range of the foundation underside, the additional stress can be calculated by using Eq. (3.5.4) and the principle of superposition.

(1) The point is in the side of the foundation

Consider a vertical uniform load p acting on a rectangular foundation underside $bhfc$, as shown in Fig. 3.5.4(a). In order to find the additional stress σ_z at any depth z under point M', we can draw one auxiliary line eM' parallel to the side of the foundation. Point M' is the common corner of two rectangles $hbM'e$ (I) and $eM'fc$(II). Therefore, the additional stress at any depth z under M' is the sum of the additional stress of the above two new foundation undersides, namely.

$$\sigma_z = (\alpha_{cI} + \alpha_{cII})p$$

(2) The point is inside of the foundation

Consider a vertical uniform load p acting on a rectangular foundation underside $abcd$, as shown in Fig. 3.5.4(b). In order to find the additional stress σ_z at any depth z under point M', we can draw two auxiliary lines eg and hf parallel to the longer side and the shorter side of the foundation respectively. Point M' is the common corner of four rectangles $bhM'e$(I), $eM'fc$(II), $hagM'$(III), and $M'gdf$(IV). Therefore, the additional stress at any depth z under M' is the sum of the additional stress of the above four new foundation undersides, namely.

$$\sigma_z = (\alpha_{cI} + \alpha_{cII} + \alpha_{cIII} + \alpha_{cIV})p$$

(3) The point is outside of the side of the foundation

Consider a vertical uniform load p acting on a rectangular foundation underside $abcd$, as shown in Fig. 3.5.4(c). In order to find the additional stress σ_z at any depth z under point M', we can draw two auxiliary lines eM' and hf parallel to the longer side and the shorter side of the foundation respectively. Point M' is the common corner of four rectangles $bhM'e$(I), $eM'fc$(II), $hagM'$(III), and $M'gdf$(IV). Therefore, the additional stress at any depth z under M' is the sum of the additional stress of the above four new foundation undersides, namely.

$$\sigma_z = (\alpha_{cI} + \alpha_{cII} - \alpha_{cIII} - \alpha_{cIV})p$$

(4) The point is outside of the corner of the foundation

Consider a vertical uniform load p acting on a rectangular foundation underside $abcd$, as shown in Fig. 3.5.4(d). In order to find the additional stress σ_z at any depth z under point M', we can draw two auxiliary lines eM' and hM'

parallel to the longer side and the shorter side of the foundation respectively. Point M' is the common corner of four rectangles $bhM'e(\text{I})$, $eM'fc(\text{II})$, $hagM'(\text{III})$, and $M'fdg(\text{IV})$. Therefore, the additional stress at any depth z under M' is the sum of the additional stress of the above four new foundation undersides, namely

$$\sigma_z = (\alpha_{cI} - \alpha_{cII} - \alpha_{cIII} + \alpha_{cIV})p$$

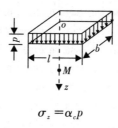

Fig. 3.5.4 Calculation point inside or outside of the foundation underside

l is the longer side, b is the shorter side (see the following figure) in Table 3.5.2.

$$\sigma_z = \alpha_c p$$

Table 3.5.2 Additional stress coefficient α_c under the corner of rectangular foundation underside due to vertical load

z/b \ l/b	1.0	1.2	1.4	1.6	1.8	2.0	3.0	4.0	5.0	6.0	10.0
0.0	0.2500	0.2500	0.2500	0.2500	0.2500	0.2500	0.2500	0.250	0.2500	0.2500	0.2500
0.2	0.2486	0.2489	0.2490	0.2491	0.2491	0.2491	0.2492	0.2492	0.2492	0.2492	0.2492
0.4	0.2401	0.2420	0.2429	0.2434	0.2437	0.2439	0.2442	0.2443	0.2443	0.2443	0.2443
0.6	0.2229	0.2275	0.2300	0.2315	0.2324	0.2329	0.2339	0.2341	0.2342	0.2342	0.2342
0.8	0.1999	0.2075	0.2120	0.2147	0.2165	0.2176	0.2196	0.2200	0.2202	0.2202	0.2202
1.0	0.1752	0.1851	0.1911	0.1955	0.1981	0.1999	0.2034	0.2042	0.2044	0.2045	0.2046
1.2	0.1516	0.1626	0.1705	0.1758	0.1793	0.1818	0.1870	0.1882	0.1885	0.1887	0.1888
1.4	0.1308	01423	0.1508	0.1569	0.1613	0.1644	0.1712	0.1730	0.1735	0.1738	0.1740
1.6	0.1123	0.1241	0.1329	0.1396	0.1445	0.1482	0.1567	0.1590	0.1598	0.1601	0.1604
1.8	0.0969	0.1083	0.1172	0.1241	0.1294	0.1334	0.1434	0.1463	0.1474	0.1478	0.1482
2.0	0.0840	0.0947	0.1034	0.1103	0.1158	0.1202	0.1314	0.1350	0.1363	0.1368	0.1374
2.2	0.0732	0.0832	0.0917	0.0984	0.1039	0.1084	0.1205	0.1248	0.1264	0.1271	0.1277
2.4	0.0642	0.0734	0.0813	0.0879	0.0934	0.0979	0.1108	0.1156	0.1175	0.1184	0.1192
2.6	0.0566	0.0651	0.0725	0.0788	0.0842	0.0887	0.1020	0.1073	0.1095	0.1106	0.1116
2.8	0.0502	0.0580	0.0649	0.0709	0.0761	0.0805	0.0942	0.0999	0.1024	0.1036	0.1048
3.0	0.0447	0.0519	0.0583	0.0640	0.0690	0.0732	0.0870	0.0931	0.0959	0.0973	0.0987
3.2	0.0401	0.0467	0.0526	0.0580	0.0627	0.0668	0.0806	0.0870	0.0900	0.096	0.0933
3.4	0.0361	0.0421	0.0477	0.0527	0.0571	0.0611	0.0747	0.0814	0.847	0.0864	0.0882
3.6	0.0326	0.0382	0.0433	0.0480	0.0523	0.0561	0.0694	0.0763	0.0799	0.0816	0.0837
3.8	0.0296	0.0348	0.0395	0.0439	0.0479	0.0516	0.0646	0.0717	0.0753	0.0773	0.0796
4.0	0.0270	0.0318	0.0362	0.0404	0.0441	0.0474	0.0603	0.0674	0.0712	0.0733	0.0758
4.2	0.0247	0.0291	0.0333	0.0371	0.0407	0.0439	0.0563	0.0634	0.0674	0.0696	0.0724
4.4	0.0227	0.0268	0.0306	0.0343	0.0376	0.0407	0.0527	0.0597	0.0639	0.0662	0.0692
4.6	0.0209	0.0247	0.0283	0.0317	0.0348	0.0378	0.0493	0.0564	0.0606	0.0630	0.0663
4.8	0.0193	0.0229	0.0262	0.0294	0.0324	0.0352	0.0463	0.0533	0.0576	0.0601	0.0635
5.0	0.0179	0.0212	0.0243	0.0274	0.0302	0.0328	0.0435	0.0504	0.0547	0.0573	0.0610
6.0	0.0127	0.0151	0.0174	0.0196	0.02218	0.0238	0.0325	0.0388	0.0431	0.0460	0.0506
7.0	0.0094	0.0112	0.0130	0.0147	0.0164	0.0180	0.0251	0.0306	0.0346	0.0376	0.0428
8.0	0.0073	0.0087	0.0101	0.0114	0.0127	0.0140	0.0198	0.0246	0.0283	0.0311	0.0367
9.0	0.0058	0.0069	0.0080	0.0091	0.0102	0.0112	0.0161	0.0202	0.0235	0.0262	0.0319
10.0	0.0047	0.0056	0.0065	0.0074	0.0083	0.0092	0.0132	0.0167	0.0198	0.0222	0.0280

Example 3.3

There are two adjacent foundations A and B, their sizes, positions and the additional stress distributions are all shown in Fig. 3.5.5. Considering the effect of the adjacent foundation B, try to find the additional stress at a depth z of 2m under the center point O of foundation A.

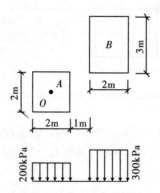

Fig. 3.5.5 Diagram work example 3.3

 olution

Step 1 The additional stress of the center point O due to the vertical uniform load of the foundation A

In order to obtain the additional stress under point O, the foundation A underside is divided into four rectangles of equal area of 1m ×1m. The additional stresses sum of the four rectangles is the additional stress of the center point O of the foundation A, namely

$$\sigma_z = 4\alpha_c p_A$$

With $l/b = 1/1 = 1$ and $z/b = 2/1 = 2$, and from Table 3.5.2, we can get $\alpha_c = 0.0840$. Therefore, the additional under point O is

$$\sigma_z = 4\alpha_c p_A = 4 \times 0.0840 \times 200 = 67.2 \text{kPa}$$

Step 2 The additional stress of the center point O due to the vertical uniform load of the foundation B

According to the point O is outside the foundation B to calculate the addition stress, as shown in Fig. 3.5.6, that is

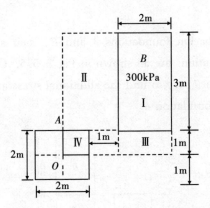

Fig. 3.5.6 The additional stress of the center point O of the foundation A

$$\sigma_z = (\alpha_{cI} - \alpha_{cII} - \alpha_{cIII} + \alpha_{cIV}) p_B$$

rectangle I : $l=4$m, $b=4$m; with $l/b=1$, $z/b=0.5$, from Table 3.5.2, we can get $\alpha_c = 0.2315$

rectangle II : $l=4$m, $b=2$m; with $l/b=2$, $z/b=1$, from Table 3.5.2, we can get $\alpha_c = 0.1999$

rectangle III : $l=4$m, $b=1$m; with $l/b=4$, $z/b=2$, from Table 3.5.2, we can get $\alpha_c = 0.1350$

rectangle IV : $l=2$m, $b=1$m; with $l/b=2$, $z/b=2$, from Table 3.5.2, we can get $\alpha_c = 0.1202$

$$\sigma_z = (\alpha_{cI} - \alpha_{cII} - \alpha_{cIII} + \alpha_{cIV}) p_B = (0.2315 - 0.1999 - 0.1350 + 0.1202) \times 300 = 5.28 \text{kPa}$$

Step 3 The additional stress of the center point O at the 2m depth of the foundation A

$$\sigma_z = 67.2 + 5.28 = 72.48 \text{kPa}$$

3.5.3 Additional stress under the corners of a rectangular foundation underside due to a vertical triangular load

When a triangularly distributed load (i.e., the triangularly distributed compressive pressure, ditto) is applied to a rectangular foundation underside, the additional stress under the corner where the load intensity is zero can also be calculated by integrating Eq. (3.5.2) with respect to the whole loading area. If the maximum intensity of the triangular load on the rectangular foundation underside is p_t, the acting force dF on the infinitesimal area $dxdy$ equals $\dfrac{p_t x}{b} dxdy$, which can be taken granted as a concentrated point load, as shown in Fig. 3.5.7. Therefore, the additional stress at an arbitrary depth z under the corner O induced by this point load can be calculated as given by

$$d\sigma_z = \frac{3p_t}{2\pi b} \cdot \frac{1}{\left[1+\left(\dfrac{r}{z}\right)^2\right]^{5/2}} \cdot \frac{xdxdy}{z^2} \qquad (3.5.5)$$

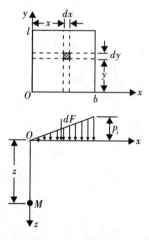

Fig. 3.5.7 Rectangular foundation subjected to triangularly distributed load

Substituting $r^2 = x^2 + y^2$ into the above equation and integrating it with respect to the whole area of the rectangular foundation underside yields the additional stress at a depth under the corner of the rectangular foundation subjected to the vertical triangular load, as given by

$$\sigma_z = \alpha_{tc} p_t \qquad (3.5.6)$$

$$\alpha_{tc} = \frac{mn}{2\pi} \left[\frac{1}{\sqrt{m^2+n^2}} - \frac{n^2}{(1+n^2)\sqrt{1+m^2+n^2}} \right] \qquad (3.5.7)$$

where α_{tc} is the coefficient of the additional stress under the corner of the underside of a rectangular foundation due to vertical triangular load. It is a function of $m(=l/b)$ and $n(=z/b)$ and it can be read off in Table 3.5.3, where b is the length of one side of the foundation underside in the loading variation direction and l is the length of another side.

For the additional stress at any point inside and outside the range of the foundation underside, it can also be calculated using the principle of superposition. Two points should, however, be noted: the calculation point should be on the vertical line under the point where the triangular load intensity is zero; b represents always the side length of the rectangular foundation underside in the loading variation direction.

The following figure is corresponding to Table 3.5.3.

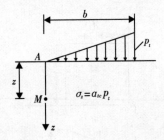

Table 3.5.3 Additional stress coefficient α_{tc} under the corner of rectangular foundation underside due to vertical triangular load

l/b \ z/b	0.2	0.4	0.6	0.8	1.0	1.2	1.4	1.6
0.0	0.0000	0.0000	0.0000	0.0000	0.0000	0.0000	0.0000	0.0000
0.2	0.0223	0.0280	0.0296	0.0301	0.0304	0.0305	0.0305	0.0306
0.4	0.0269	0.0420	0.0487	0.0517	0.0531	0.0539	0.0543	0.0545
0.6	0.0259	0.0448	0.0560	0.0621	0.0654	0.0673	0.0684	0.0690
0.8	0.0232	0.0421	0.0553	0.0637	0.0688	0.0720	0.0739	0.0751
1.0	0.0201	0.0375	0.0508	0.0602	0.0666	0.0708	0.0735	0.0753
1.2	0.0171	0.0324	0.0450	0.0546	0.0615	0.0664	0.0698	0.0721
1.4	0.0145	0.0278	0.0392	0.0483	0.0554	0.0606	0.0644	0.0672
1.6	0.0123	0.0238	0.0339	0.0424	0.0492	0.0545	0.0586	0.0616
1.8	0.0105	0.0204	0.0294	0.0371	0.0435	0.0487	0.0528	0.0560
2.0	0.0090	0.0176	0.0255	0.0324	0.0384	0.0434	0.0474	0.0507
2.5	0.0063	0.0125	0.0183	0.0236	0.0284	0.0326	0.0362	0.0393
3.0	0.0046	0.0092	0.0135	0.0176	0.0214	0.0249	0.0280	0.0307
5.0	0.0018	0.0036	0.0054	0.0071	0.0088	0.0104	0.0120	0.0135
7.0	0.0009	0.0019	0.0028	0.0038	0.0047	0.0056	0.0064	0.0073
10.0	0.0005	0.0009	0.0014	0.0019	0.0023	0.0028	0.0033	0.0037

to be continued

l/b \ z/b	1.8	2.0	3.0	4.0	6.0	8.0	10.0
0.0	0.0000	0.0000	0.0000	0.0000	0.0000	0.0000	0.0000
0.2	0.0306	0.0306	0.0306	0.0306	0.0306	0.0306	0.0306
0.4	0.0546	0.0547	0.0548	0.0549	0.0549	0.0549	0.0549
0.6	0.0694	0.0696	0.0701	0.0702	0.0702	0.0702	0.0702
0.8	0.0759	0.0764	0.0773	0.0776	0.0776	0.0776	0.0776
1.0	0.0766	0.0774	0.0790	0.0794	0.0795	0.0796	0.0796
1.2	0.0738	0.0749	0.0774	0.0779	0.0782	0.0783	0.0783
1.4	0.0692	0.0707	0.0739	0.0748	0.0752	0.0752	0.0753
1.6	0.0639	0.0656	0.0697	0.0708	0.0714	0.0715	0.0715
1.8	0.0585	0.0604	0.0652	0.0666	0.0673	0.0675	0.0675
2.0	0.0533	0.0553	0.0607	0.0624	0.0634	0.0636	0.0636
2.5	0.0419	0.0440	0.0504	0.0529	0.0543	0.0547	0.0548
3.0	0.0331	0.0352	0.0419	0.0449	0.0469	0.0474	0.0476
5.0	0.0148	0.0161	0.0214	0.0248	0.0283	0.0296	0.0301
7.0	0.0081	0.0089	0.0124	0.0152	0.01860	0.0204	0.0212
10.0	0.0041	0.0046	0.0066	0.0084	0.0111	0.0128	0.0139

3.5.4 Additional stress under corners of rectangular foundation underside due to horizontal load

When a horizontal uniform load p_h is applied to a rectangular foundation underside (Fig. 3.5.8), the additional stress at an arbitrary depth z under the corner of the foundation is

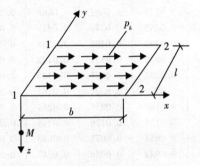

Fig. 3.5.8 Rectangular foundation subjected to horizontal uniform load

$$\sigma_z = \pm \alpha_l \cdot p_h \qquad (3.5.8a)$$

$$\alpha_l = \frac{m}{2\pi}\left[\frac{1}{\sqrt{m^2+n^2}} - \frac{n^2}{(1+n^2)\sqrt{1+m^2+n^2}}\right] \qquad (3.5.8b)$$

where α_l is the coefficient of the additional stress under the corner of the underside of a rectangular foundation due to horizontal uniform load. It is a function of $m(=l/b)$ and $n(=z/b)$ and it can be read off in Table 3.5.4, where b is the length of one side of the foundation underside in the direction parallel to the horizontal load direction and l is the length of another side.

In the above equation, the positive sign '+' is taken at the calculation point under the stopping end of the horizontal uniform load (under point 2) and the negative sign '−' is taken at the calculation point under the starting end of the horizontal uniform load (under point 1).

For the additional stress at any point inside and outside the range of the foundation underside, it can also be calculated using the principle of superposition.

Table 3.5.4 Additional stress coefficient α_t under the corner of rectangular foundation underside due to horizontal uniform load

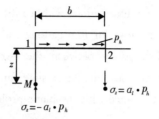

l/b z/b	0.2	0.4	1.0	1.6	2.0	3.0	4.0	6.0	8.0	10.0
0.0	0.1592	0.1592	0.1592	0.1592	0.1592	0.1592	0.1592	0.1592	0.1592	0.1592
0.2	0.1114	0.1401	0.1518	0.1528	0.1529	0.1530	0.1530	0.1530	0.1530	0.1530
0.4	0.0672	0.1049	0.1328	0.1362	0.1367	0.1371	0.1372	0.1372	0.1372	0.1372
0.6	0.0432	0.0746	0.1091	0.1150	0.1160	0.1168	0.1169	0.1170	0.1170	0.1170
0.8	0.0290	0.0527	0.0861	0.0939	0.0955	0.0967	0.0969	0.0970	0.0970	0.0970
1.0	0.0201	0.0375	0.0666	0.0753	0.0774	0.0790	0.0794	0.0795	0.0796	0.0796
1.2	0.0142	0.0270	0.0512	0.0601	0.0624	0.0645	0.0650	0.0652	0.0652	0.0652
1.4	0.0103	0.0199	0.0395	0.0480	0.0505	0.0528	0.0534	0.0537	0.0537	0.0538
1.6	0.0077	0.0149	0.0308	0.0385	0.0410	0.0436	0.0443	0.0446	0.0447	0.0447
1.8	0.0058	0.0113	0.0242	0.0311	0.0336	0.0362	0.0370	0.0374	0.0375	0.0375
2.0	0.0045	0.0088	0.0192	0.0253	0.0277	0.0303	0.0312	0.0317	0.0318	0.0318
2.5	0.0025	0.0050	0.0113	0.0157	0.0176	0.0202	0.0211	0.0217	0.0219	0.0219
3.0	0.0015	0.0031	0.0071	0.0102	0.0117	0.0140	0.0150	0.0156	0.0158	0.0159
5.0	0.0004	0.0007	0.0018	0.0027	0.0032	0.0043	0.0050	0.0057	0.0059	0.0062
7.0	0.0001	0.0003	0.0007	0.0010	0.0013	0.0018	0.0022	0.0027	0.0029	0.0030
10.0	0.0001	0.0001	0.0002	0.0004	0.0005	0.0007	0.0008	0.0011	0.0013	0.0014

3.6 Additional stress in plane problem

In theory, when the length l to width b ratio of a foundation approaches an infinite large value, the stress state in the interior of the ground is a plane problem. In engineering practice, however, there is no infinitely long foundation in reality. Based on researches, the stress state for foundations with a ratio l/b greater than or equal to 10 is more or less the same as those for foundations with a ratio l/b approaching an infinite large value. The discrepancy between the two cases is acceptable in an engineering point of view. Sometimes, the stress state for foundations with l/b greater than 5 is also calculated in accordance with the plane problem.

◀ 3.6.1　Stress increase due to vertical line load

Vertical line load is defined as the vertical uniform load applied on an infinitely long line, as shown in Fig. 3.6.1. When a vertical line load is applied on the ground surface, the stress increase at an arbitrary depth z in the interior of the ground can be calculated using the Flamant solution, as given by

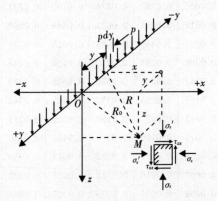

Fig. 3.6.1　Stress state of the soil due to vertical line load

$$\sigma_z = \frac{2pz^3}{\pi(x^2+z^2)^2} \tag{3.6.1a}$$

$$\sigma_x = \frac{2px^2 z}{\pi(x^2+z^2)^2} \tag{3.6.1b}$$

$$\tau_{zx} = \tau_{xz} = \frac{2pxz^2}{\pi(x^2+z^2)^2} \tag{3.6.1c}$$

where p denotes the line load on a unit length (kN/m); x and z are coordinates of the calculation point.

3.6.2 Additional stress under a strip foundation underside due to vertical uniform load

When a vertical uniform load of intensity p is applied to the foundation underside as shown in Fig. 3.6.2, the additional stress at an arbitrary point M induced by the line load $\overline{dp} = pd\xi$ on an infinitesimal width $d\xi$ can firstly be calculated by using Eq. (3.6.2), as given by

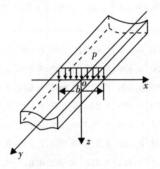

Fig. 3.6.2 Strip foundation subjected to vertical uniform load

$$d\sigma_z = \frac{2p}{\pi} \cdot \frac{z^3 d\xi}{[(x-\xi)^2 + z^2]^2} \quad (3.6.2)$$

The additional stress under the strip foundation underside due to vertical uniform load can then be calculated by integrating the above equation with respect to the width of the loading area, as expressed by

$$\sigma_z = \int_0^b \frac{2p}{\pi} \cdot \frac{z^3 d\xi}{[(x-\xi)^2 + z^2]^2}$$

$$= \frac{p}{\pi} \left[\arctan\left(\frac{m}{n}\right) - \arctan\left(\frac{m-1}{n}\right) + \frac{mn}{n^2 + m^2} - \frac{n(m-1)}{n^2 + (m-1)^2} \right] \quad (3.6.3)$$

$$= \alpha_z^s p$$

where α_z^s is the coefficient of the additional stress under the underside of a strip foundation due to vertical uniform load (compressive pressure, ditto). It is a function of m ($=x/b$) and n (z/b) and it can be read off in Table 3.6.1, where b is the width of the strip foundation as shown in Fig. 3.6.2; x and z are coordinates of the calculation point.

Table 3.6.1 Additional stress coefficient α_z^s under the underside of a strip foundation due to vertical uniform load

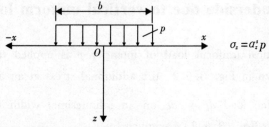

$\sigma_z = \alpha_z^s p$

x/b \ z/b	0.0	0.10	0.25	0.35	0.50	0.75	1.00	1.50	2.00	2.50	3.00	4.00	5.00
0.00	1.000	1.000	1.000	1.000	0.500	0.000	0.000	0.000	0.000	0.000	0.000	0.000	0.000
0.05	1.000	1.000	0.995	0.970	0.500	0.002	0.000	0.000	0.000	0.000	0.000	0.000	0.000
0.10	0.997	0.996	0.986	0.965	0.499	0.010	0.005	0.000	0.000	0.000	0.000	0.000	0.000
0.15	0.993	0.987	0.968	0.910	0.498	0.033	0.008	0.001	0.000	0.000	0.000	0.000	0.000
0.25	0.960	0.954	0.905	0.805	0.496	0.088	0.019	0.002	0.001	0.000	0.000	0.000	0.000
0.35	0.907	0.900	0.832	0.732	0.492	0.148	0.039	0.006	0.003	0.001	0.000	0.000	0.000
0.50	0.820	0.812	0.735	0.651	0.481	0.218	0.082	0.017	0.005	0.002	0.001	0.000	0.000
0.75	0.668	0.658	0.610	0.552	0.450	0.263	0.146	0.040	0.017	0.005	0.005	0.001	0.000
1.00	0.552	0.541	0.513	0.475	0.410	0.288	0.185	0.071	0.029	0.013	0.007	0.002	0.001
1.50	0.396	0.395	0.379	0.353	0.332	0.273	0.211	0.114	0.055	0.030	0.018	0.006	0.003
2.00	0.306	0.304	0.292	0.288	0.275	0.242	0.205	0.134	0.083	0.051	0.028	0.013	0.006
2.50	0.245	0.244	0.239	0.237	0.231	0.215	0.188	0.139	0.098	0.065	0.034	0.021	0.010
3.00	0.208	0.208	0.206	0.202	0.198	0.185	0.171	0.136	0.103	0.075	0.053	0.028	0.015
4.00	0.160	0.160	0.158	0.156	0.153	0.147	0.140	0.122	0.102	0.081	0.066	0.040	0.025
5.00	0.126	0.126	0.125	0.125	0.124	0.121	0.117	0.107	0.095	0.082	0.069	0.046	0.034

3.6.3 Additional stress under a strip foundation underside due to vertical triangular load

When a triangular distributed load with a maximum intensity of p_t is applied to a strip foundation underside as shown in Fig. 3.6.3, the additional stress at a calculation point M induced by the vertical line load $dp = \frac{p_t}{b}\xi d\xi$ on an infinitesimal width $d\xi$ can firstly be calculated by using Eq. (3.6.1a). By integrating this additional stress on $d\xi$ with respect to the width of the foundation, the additional stress at point M induced by the vertical triangular load can then be calculated, as given by

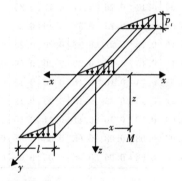

Fig. 3.6.3 Strip foundation subjected to vertical triangular distributed load

$$\sigma_z = \int_0^b \frac{2p_t}{\pi b} \cdot \frac{z^3 \xi d\xi}{[(x-\xi)^2 + z^2]^2}$$

$$= \frac{p_t}{\pi}\left\{m\left[\arctan\left(\frac{m}{n}\right) - \arctan\left(\frac{m-1}{n}\right)\right] - \frac{n(m-1)}{n^2 + (m-1)^2}\right\} \quad (3.6.4)$$

$$= \alpha_t^s p_t$$

where α_t^s is the coefficient of the additional stress under the underside of a strip foundation due to vertical triangular distributed load (compressive pressure, ditto). It is a function of m ($=x/b$) and n (z/b) and it can be read off in Table 3.6.2, where b is the width of the strip foundation as shown in Fig. 3.6.3. The other symbols have the same meanings as above.

Table 3.6.2 Additional stress coefficient α_z^t under the underside of a strip foundation due to vertical triangular distributed load

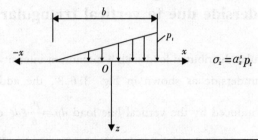

$\sigma_z = \alpha_z^t p_t$

x/b z/b	-2.00	-1.50	-1.00	-0.75	-0.50	-0.25	0.00	0.25	0.50	0.75	1.00	1.50	2.00	3.00
0.00	0.00	0.00	0.00	0.00	0.00	0.00	0.25	0.50	0.75	0.50	0.00	0.00	0.00	0.00
0.25	0.00	0.00	0.00	0.01	0.08	0.26	0.48	0.65	0.42	0.08	0.02	0.00	0.00	0.00
0.50	0.00	0.01	0.02	0.05	0.13	0.26	0.41	0.47	0.35	0.16	0.06	0.01	0.00	0.00
0.75	0.01	0.01	0.05	0.08	0.15	0.25	0.33	0.36	0.29	0.19	0.10	0.03	0.01	0.00
1.00	0.01	0.03	0.06	0.10	0.16	0.22	0.28	0.29	0.25	0.18	0.12	0.05	0.02	0.00
1.50	0.02	0.05	0.09	0.11	0.15	0.18	0.20	0.20	0.19	0.16	0.13	0.07	0.04	0.01
2.00	0.03	0.06	0.09	0.11	0.14	0.16	0.16	0.16	0.15	0.13	0.12	0.08	0.05	0.02
2.50	0.04	0.06	0.08	0.12	0.13	0.13	0.13	0.13	0.12	0.11	0.10	0.07	0.05	0.02
3.00	0.05	0.06	0.08	0.09	0.10	0.10	0.11	0.11	0.10	0.10	0.09	0.07	0.05	0.03
4.00	0.05	0.06	0.07	0.07	0.08	0.08	0.08	0.08	0.08	0.08	0.07	0.06	0.05	0.03
5.00	0.05	0.05	0.06	0.06	0.06	0.06	0.06	0.06	0.06	0.06	0.06	0.05	0.04	0.03

3.6.4 Additional stress under a strip foundation underside due to horizontal uniform load

Similarly, when a horizontal uniform load with an intensity of p_h is applied to a strip foundation underside as shown in Fig. 3.6.4, the additional stress at an arbitrary calculation point M can be calculated by integrating the additional stress due to a horizontal line load with respect of the width of the foundation, as given by

Chapter 3 Stress distribution in soils

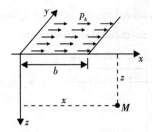

Fig. 3.6.4 Strip foundation subjected to horizontal uniform load

$$\sigma_z = \frac{p_h}{\pi} \left[\frac{n^2}{(m-1)^2 + n^2} - \frac{n^2}{m^2 + n^2} \right]$$
$$= \alpha_h p \tag{3.6.5}$$

where α_h is the coefficient of the additional stress under the underside of a strip foundation due to horizontal uniform load (compressive pressure, ditto). It is a function of m ($=x/b$) and n (z/b) and it can be read off in Table 3.6.3. The other symbols have the same meanings as above.

It must be noted that when calculating the additional stress under the strip foundation, the coordinate system must satisfy the requirements specified separately in Figs. 3.6.1, 3.6.2 and 3.6.3.

Table 3.6.3 Additional stress coefficient α_h under the underside of a strip foundation due to horizontal uniform load

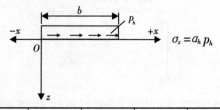

x/b \ z/b	0.00	0.25	0.50	0.75	1.00	1.25	1.50	-0.25
0.01	-0.318	-0.001	0.000	0.001	0.318	0.001	0.001	-0.001
0.1	-0.315	-0.039	0.000	0.039	0.315	0.042	0.011	-0.042
0.2	-0.306	-0.103	0.000	0.103	0.306	0.116	0.038	-0.116
0.4	-0.274	-0.159	0.000	0.159	0.274	0.199	0.103	-0.199
0.6	-0.234	-0.147	0.000	0.147	0.234	0.212	0.144	-0.212
0.8	-0.194	-0.121	0.000	0.121	0.194	0.197	0.158	-0.197
1.0	-0.159	-0.096	0.000	0.096	0.159	0.175	0.157	-0.175
1.2	-0.131	-0.078	0.000	0.078	0.131	0.153	0.147	-0.153
1.4	-0.108	-0.061	0.000	0.061	0.108	0.132	0.133	-0.132
2.0	-0.064	-0.034	0.000	0.034	0.064	0.085	0.096	-0.085

Example 3.4

A concentrated load $F = 400\text{kN/m}$ and a moment $M = 20\text{kN} \cdot \text{m}$ are acting on the strip foundation, as shown in Fig. 3.6.5. Try to find the additional stress of the center of the strip foundation and drawing its profile with the depth of soil.

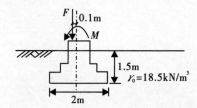

Fig. 3.6.5 Diagram work example 3.4

Chapter 3 Stress distribution in soils

olution

Step 1 Find the offsetting e

$$e = \frac{M + F \times 0.1}{F + G} = \frac{20 + 400 \times 0.1}{400 + 20 \times 2 \times 1 \times 1.5} = 0.13 m$$

Step 2 Find the contact pressure

$$p_{\substack{max \\ min}} = \frac{F+G}{b}(1 \pm \frac{6e}{b}) = \frac{400+60}{2}(1 \pm \frac{6 \times 0.13}{2})$$

$$= \frac{319.7 \text{kPa}}{140.3 \text{kPa}}$$

The contact pressure diagram is shown as Fig. 3.6.6.

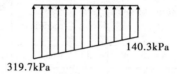

Fig. 3.6.6 Contact pressure diagram

Step 3 Find the additional stress of the foundation underside

$$p_{\substack{0max \\ 0min}} = p_{\substack{max \\ min}} - \gamma_0 d = \frac{319.7}{140.3} - 18.5 \times 1.5 = \frac{292.0 \text{kPa}}{112.6 \text{kPa}}$$

The additional stress of the foundation underside diagram is shown as Fig. 3.6.7.

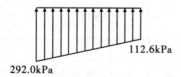

Fig. 3.6.7 Additional stress of the foundation underside diagram

Step 4 Find the additional stress at the center of the foundation

The trapezoidal vertical additional stress of the foundation underside is decomposed into a vertical uniform additional stress $p_1 = 112.6$ kPa and a vertical triangular additional stress $p_{2t} = 179.4$ kPa, as shown in Fig. 3.6.8. The additional stress of the center of the foundation due to each load can be calculated and the calculation process is shown in following table, the additional stress coefficients of the table 3.6.4 are read off in Table 3.6.1 and 3.6.2.

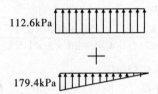

Fig. 3.6.8 Additional stress of the foundation underside breakdown drawing

Table 3.6.4 Coefficients of additional stress

Type of load	x(m)	z(m)	b(m)	x/b	z/b	Additional stress coefficient	Additional stress/kPa	Sum of additional stress due to the two loads
Vertical uniform load	0	0	2	0	0	1.00 (α'_z)	112.6	202.3
	0	0.5	2	0	0.25	0.960	108.096	194.2
	0	1	2	0	0.5	0.820	92.332	165.9
	0	2	2	0	1	0.552	62.155	112.4
	0	3	2	0	1.5	0.396	44.59	80.5
	0	4	2	0	2	0.306	34.456	63.2
Vertical triangular distributed load	0	0	2	0	0	0.50(α'_t)	89.7	
	0	0.5	2	0	0.25	0.48	86.112	
	0	1	2	0	0.5	0.41	73.554	
	0	2	2	0	1	0.28	50.232	
	0	3	2	0	1.5	0.20	35.88	
	0	4	2	0	2	0.16	28.704	

$$\sigma_z = \alpha^s_z p_1 + \alpha^s_t p_{2t}$$

The additional stress of the center of the foundation diagram is shown in Fig. 3.6.9.

Chapter 3   Stress distribution in soils

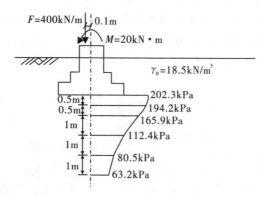

Fig. 3.6.9   Additional stress of the center of the foundation diagram

EXERCISES

3.1   A ground soil profile is shown in Fig. 3.1. Find the vertical geostatic stress and plot its distribution curve.

3.2   A rectangular foundation is shown in Fig. 3.2. A vertical load of 800kN is applied on point A of the foundation. If the eccentricity is 0.5m, try to find the additional stress at a depth 5m under point B in the figure.

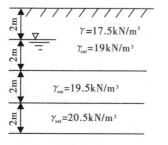

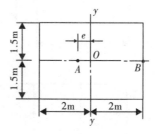

Fig. 3.1   Exercise 3.1   Fig. 3.2   Exercise 3.2

3.3   The two foundation sizes are shown in Fig. 3.3. A vertical centric load 1940kN is applied on the foundation I and II. The embedded depths of the two foundations are all 1.5m. Find the additional stress distribution under the center of the foundation I, with consideration of the influence of the adjoining foundation (the calculation depth is $z=5b$, the unit weight of soil upwards the floor is 18kN/m^3).

3.4 The foundation is shown in Fig. 3.4. A uniform stress p is applied on the foundation Find the additional stress at the depth of 10m under the point A and O respectively.

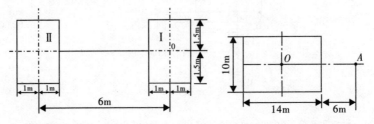

Fig. 3.3　Exercise 3.3　　　Fig. 3.4　Exercise 3.4

3.5 The foundation is shown in Fig. 3.5. Find the additional stress distribution under the point O and A of the foundation (the calculation depth is $3b$).

3.6 The strip foundation is shown in Fig. 3.6. Find the additional stress distribution of the point A.

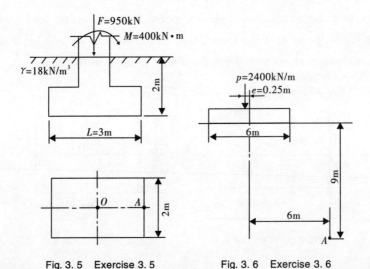

Fig. 3.5　Exercise 3.5　　　Fig. 3.6　Exercise 3.6

REFERENCES

1. Soil mechanics work team at Hohai University. Soil Mechanics [M]. China Communication Press, Beijing, 2004.
2. Siwei He. Essentials of Soil Mechanics [M]. Zhongshan University Press,

Guangzhou, 2003.
3. Ministry of Construction P. R. China. GB50021 −2001 Code for Investigation of Geotechnical Engineering (2009 revised edition) [S]. Beijing: China Building Industry Press, 2009.
4. Ministry of Construction P. R. China. GB50025 − 2007 Code for Building Construction in Collapsible Loess Zone[S]. Beijing: China Building Industry Press, 2004.
5. Ministry of Water Resources P. R. China. GB/T50145 −2007 Standard for Engineering Classification of Soil[S]. Beijing: China Planning Press, 2008.
6. P. R. China Ministry of Housing and Urban-Rural Development Producer. GB50007 −2011 Code for Design of Building Foundation[S]. Beijing: China Building Industry Press, 2009.
7. Nanjing Hydraulic Research Institute. SL237 −1999 Specification of Soil Test [S]. Beijing: China Waterpower Press, 1999.
8. Ministry of Water Resources P. R. China. GB/T 50123 −1999 Standard for Soil Test Method[S]. Beijing: China Planning Press, 2000.
9. Shude Zhao, Hongjian Liao. Soil Mechanics (the second edition) [M]. Beijing: Higher Education Press, 2010.
10. Shude Zhao, Hongjian Liao. Civil Engineering Geology[M]. Beijing: Science Press, 2009.

SOIL MECHANICS
CHAPTER 4

Compression and consolidation of soils

4.1 Introduction

There are two kinds of stress in soils, i.e., gravity stress and additional stress. The deformation of soils is mainly caused by the additional stress. When the soil layers are subjected to the loading from buildings, the deformation occurs. The settlement is small as the soil is hard, which has no influence on the usage of engineering. However, as the soil is soft and its thickness is uneven, or the weight of building varies sharply, the large settlement occurs, which will cause the various accidents and affect the usage of engineering.

For some special soils, the deformation also occurs due to the change of water content. For collapsible loess, the additional settlement that is caused by the increase of water content is called as collapsible settlement. For the regional expansive soil, the increase of water content can induce the expansion of ground, even the crack of buildings.

The foundations will produce differential settlement if the vertical deformation is uneven. The large settlement or differential settlement can affect the usage of buildings and even cause failure. In order to guarantee the safety and usage of building, we must estimate the maximum settlement or differential settlement of foundation.

4.2 Compressibility characteristics

◀ 4.2.1 Compression of soils

The settlement is mainly dependent on the compressibility of soils and the loading. The reasons that induce settlement are listed as follows:

(1) External influencing factors
① The loading from buildings;
② The drop of ground water table;
③ The influence of construction;
④ The influence of vibrations;
⑤ The change of temperature.

(2) Internal influencing factors
① The compressibility of solid phase;
② The compressibility of liquid phase;
③ The compressibility of void.

For the external factors, the loading from buildings is the dominant factor, which causes the compression of voids of soil. The process of the increase of the compression of saturated soil with time is called consolidation.

The settlement mainly includes two aspects: one is the total settlement; the other is the relationship of settlement and time. To obtain the settlement, the compressibility index must be determined, which can be obtained from the compression test of field in-situ test.

◀ 4.2.2 Oedometer test

Figure 4.2.1 is schematic of Oedometer test. Consolidation settlement is the vertical displacement of the surface corresponding to the volume change at any stage of the consolidation process. The characteristics of a soil during one-dimensional consolidation or swelling can be determined by the oedometer test. Figure 4.2.1 shows a cross-section diagram through an oedometer. The confining ring imposes a condition of zero lateral strain on the specimen.

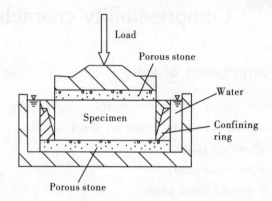

Figure 4.2.1   Oedometer test

According to the relationship of $\Delta h - p$, the curve of e-p is obtained, which indicates the change of void under the different pressure.

Figure 4.2.2a shows the relationship of e-p. It can also be plotted in a logarithmic scale (see Fig. 4.2.2b). For the different soil types, the deformations are different. For sands, the curve of e-p is flat, which indicates the void ratio decreases slowly as the increase of pressure. For clay, the curve is steep, which means the void ratio decreases significantly as pressure increases. The results also indicate that the compressibility of clay is greater than that of sand.

Three compressibility indexes can be determined from the compression curve, i.e., compression coefficient a, compression index C_c, oedometric modulus E_s.

(1) Compression coefficient a

$$a = -\frac{de}{dp} \qquad (4.2.1)$$

where negative sign means that e decreases with the increases of p. As the change range of pressure induced by external loading is small, e.g. from p_1 to p_2 (see Fig. 4.2.3), the curve $M_1 M_2$ can be approximately regarded as linear. The slope of line is:

Chapter 4 Compression and consolidation of soils

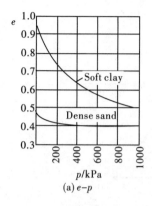
(a) $e-p$

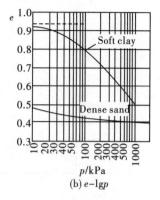
(b) $e-\lg p$

Figure 4.2.2 Relationship between void ratio and effective stress

$$a = -\frac{\Delta e}{\Delta p} = \frac{e_1 - e_2}{p_2 - p_1} \qquad (4.2.2)$$

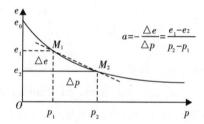

Figure 4.2.3 Curve of e-p for soil

The compression coefficient a expresses the decrease value of void ratio under unit pressure. So the larger of a, the larger of compressibility of soils is.

It should be noted that a is not a constant for a certain soil type. In order to facilitate the comparison and the application of different regions, *Code for Design of Building Foundation* (GB50007 −2011) takes a_{1-2} corresponding to $p_1 = 100\text{kPa}$ to $p_2 = 200\text{kPa}$ to evaluate the compressibility:

$$a_{1-2} < 0.1\text{MPa}^{-1} \qquad \text{low compressibility soil}$$
$$0.1\text{MPa}^{-1} \leq a_{1-2} < 0.5\text{MPa}^{-1} \qquad \text{medium compressibility soil}$$
$$a_{1-2} \geq 0.5\text{MPa}^{-1} \qquad \text{high compressibility soil}$$

(2) Compression index C_c

The compression index (C_c) is the slope of the linear portion of the e-$\lg p$ plot

• • • 105

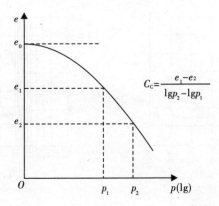

Figure 4.2.4 Calculation of C_c in e-lgp

(see Fig. 4.2.4) and is dimensionless. For any two points on the linear portion of the plot, C_c is

$$C_c = \frac{e_1 - e_2}{\lg p_2 - \lg p_1} = \frac{e_1 - e_2}{\lg \left(\frac{p_2}{p_1}\right)} \quad (4.2.3)$$

Similar to compression coefficient, the value can also judge the compressibility. The larger of C_c, the larger of void ratio is and the larger of compressibility is. Generally, as $C_c < 0.2$, soil belongs to low compressibility; $C_c = 0.2 \sim 0.4$, soil is medium compressibility, and $C_c > 0.4$, soil is high compressibility.

(3) Oedometric modulus E_s

The ratio of the incremental vertical stress σ_z to strain λ_z is called oedometric modulus E_s:

$$E_s = \frac{\sigma_z}{\lambda_z} \quad (4.2.4)$$

In above odeometer test, vertical pressure increases from p_1 to p_2, and the height decreases from h_1 to h_2, simultaneously.

Incremental stress $\sigma_z = p_2 - p_1$

Incremental strain $\lambda_z = \dfrac{h_1 - h_2}{h_1}$

Oedometric modulus E_s is

Chapter 4 Compression and consolidation of soils

$$E_s = \frac{p_2 - p_1}{h_1 - h_2} h_1 \qquad (4.2.5)$$

(4) Relationship of oedometric modulus and compression coefficient

All of oedometric modulus and compression coefficient are commonly used to express the compressibility of ground in civil engineering. They are determined from the oedometer test. So they are dependent on each other.

The compression of soil layers is schematic as Fig. 4.2.5. The area of soil sample is unit. At the start of compression, the volume of solids (i.e., soil particles) is V_s, and the volume of void is V_{v_0}. Taking $V_s = 1$, the void ratio is $e_0 = V_{v_0}$, and the total volume is $1 + e_0$ (see Fig. 4.2.5). At the end of compression, the volume of solid V_s keeps constant, and the volume reduces to V_{v_1}. The void ratio is $e_1 = V_{v_1}$ (see Fig. 4.2.5). So

$$\lambda_z = \frac{V_{v_0} - V_{v_1}}{V_{v_0}} = \frac{h_0 - h_1}{h_0} = \frac{e_0 - e_1}{1 + e_0} \qquad (4.2.6)$$

$$E_s = \frac{\sigma_z}{\lambda_z} = \frac{p_2 - p_1}{e_0 - e_1}(1 + e_0) = \frac{1 + e_0}{a} \qquad (4.2.7)$$

It can be found from Eq. (4.2.7) that E_s is inversely proportional to a. The lager of E_s, the smaller a of is, which means the compressibility is smaller. In practice, as $E_s < 4$MPa, the soil belong to high compressibility soil type, as 4MPa $\leq E_s \leq 20$MPa, the soil is medium compressibility, and as $E_s > 20$MPa, the soil is low compressibility.

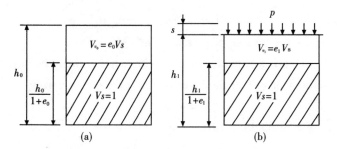

Figure 4.2.5 Schematic of compression of soil layers

The compression of soil specimen under pressure is measured by means of a dial gauge operating on the loading cap. The void ratio (e) at the end of each increment period can be calculated from the dialgauge readings (s). The phase diagram is aslo shown in Fig. 4.2.5, the method of calculation is as follows

$$h_0 A = V_{v_0} + V_s = (1 + e_0) V_s \qquad (4.2.8)$$

$$(h_0 - s) A = V_{v_1} + V_s = (1 + e_1) V_s \qquad (4.2.9)$$

$$V_s = \frac{h_0 A}{1 + e_0} = \frac{(h_0 - s) A}{1 + e_1} \qquad (4.2.10)$$

$$e_1 = e_0 - \frac{s}{h_0}(1 + e_0) \qquad (4.2.11)$$

$$s = \frac{e_0 - e_1}{1 + e_0} h_0 \qquad (4.2.12)$$

$$\varepsilon_v = \varepsilon_z = \frac{s}{h_0} = \frac{e_{10} - e_1}{1 + e_0} = \frac{\Delta e}{1 + e_0} \qquad (4.2.13)$$

where h_0 is the thickness of specimen at the start of test; e_0 is the void ratio of specimen at the start of test, which can be determined as the following equation

$$e_0 = \frac{d_s(1 + w_0)\rho_w}{\rho_0} - 1 \qquad (4.2.14)$$

h_1 and e_1, are the thickness and void ratio at end of any increment period, respectively. A is area of specimen. s is the compression of the specimen under pressure p.

The coefficient of volume compressibility m_v defined as the volume change per unit volume per unit increase in effective stress. If for an increase in effective stress from σ_0' to σ_1', the void ratio decreases from e_0 to e_1, then

$$m_v = \frac{\varepsilon_v}{\Delta \sigma'} = \frac{\Delta e}{(1 + e_0)\Delta \sigma'} = \frac{1}{1 + e_0}\left(\frac{e_0 - e_1}{\sigma_1' - \sigma_0'}\right) \qquad (4.2.15)$$

The units of m_v are the inverse of pressure (m^2/MN).

If m_v and $\Delta \sigma'$ are assumed constant with depth, then the one-dimensional consolidation settlement (s) of the layer of thickness h_0 is given by

$$s = m_v \Delta \sigma' h_0 \qquad (4.2.16)$$

or, in the case of a normally consolidated clay,

$$s = \frac{C_c \lg(\sigma_1'/\sigma_0')}{1 + e_0} h_0 \qquad (4.2.17)$$

4.3 Calculation of settlement of foundation

Settlement of buildings is induced by the deformation of soils. There are many methods to calculate the settlement of buildings, including the layer-wise summation method, code method and elastic method. Herein, only the layer-wise summation method is presented.

For the calculation of the compression of soil, the layer-wise summation method is most widely used in engineering, which is based on the formula of compression under lateral confining condition. The assumptions are as follows:

(1) The compression of soil is due to the deformation of skeleton of soil induced by the reduction of void volume. The soil particles are incompressible.

(2) Only the vertical compression occurs in soil without lateral deformation.

(3) The pressure distributes uniformly in the thickness in soil layers.

(a) Calculation principle

For the layer-wise summation method, the ground is firstly divided into several horizontal layers whose thicknesses are $h_1, h_2, h_3, \cdots, h_n$, respectively (see Fig. 4.3.1), and the compressions $s_1, s_2, s_3, \cdots, s_n$ of each soil layer are calculated, and then they all are summed together, which is the settlement of ground base,

$$s = s_1 + s_2 + \cdots + s_n = \sum_{i=1}^{n} s_i \qquad (4.3.1)$$

Calculation method and procedure for the layer-wise summation method is as follows.

(1) Drawing the profile maps of ground and foundation according to the scale.

(2) Dividing the ground into several layers. A layer surface should be kept at

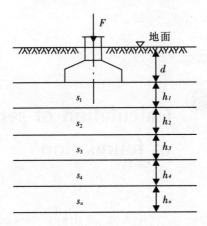

Figure 4.3.1 Principle of layer-wise summation method

the interface between the natural soil layer and underground water level. The thickness of soil layer for one type of soils should not be too larger, which is equal to or less than $0.4b$.

(3) The pressure of each layer is regarded as uniform distribution.

(4) Calculating the gravity stress and additional stress along the central axis of foundation and plot the stress envelop as shown in Fig. 4.3.2.

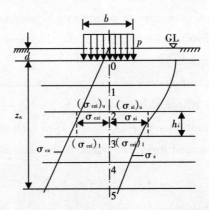

Fig. 4.3.2 Sample diagram showing one-dimensional method

It should be noted that the base pressure $p = p_0 - \gamma_0 d$ (γ_0 is weight of soil in the range of embedded depth of foundation) is used to calculate the additional pressure in ground (from the base of foundation).

(5) Determining the calculation depth of settlement. The experiences show that the depth is called the lower limit of compression layers or the calculation depth of settlement z_n when the additional stress and gravity at central axis of foundation satisfies $\sigma_z \leq 0.2\sigma_{cz}$. When there are the soft soil layers under z_n, the calculation depth should satisfy $\sigma_z \leq 0.1\sigma_{cz}$.

(6) Calculating the average gravity σ_{czi} and additional stress σ_{zi} as follows:

$$\sigma_{czi} = \frac{(\sigma_{czi})_u + (\sigma_{czi})_l}{h_i} \quad (4.3.3)$$

$$\sigma_{zi} = \frac{(\sigma_{zi})_u + (\sigma_{zi})_l}{h_i} \quad (4.3.4)$$

(7) If compression curve of e-p of ground is known, according to the initial stress and additional stress of ith layer, i.e., $p_{2i} = \overline{\sigma}_{czi} + \overline{\sigma}_{zi}$, the corresponding initial void ratio and the stable void ratio can be obtained based on the compression curve.

(8) According to $s = \dfrac{e_1 - e_2}{1 + e_1} h$, the compression of i layer can be calculated, i.e.,

$$s_i = \frac{e_{1i} - e_{2i}}{1 + e_{1i}} h_i.$$

(9) Finally, the settlement of foundation is the summation of compression of all layers, $s = \sum\limits_{i=1}^{n} s_i$.

Sometimes, if an exploration unit offers not the compression curve, but other compression indicators, such as compression coefficient, and confined compression modulus, then the formulations $s_i = \dfrac{a_i}{1 + e_{1i}} \overline{\sigma}_{zi} h_i$, $s_i = \dfrac{\overline{\sigma}_{zi}}{E_{si}} h_i$ are used to calculate the compression of each soil layer. The total settlement of foundation can be obtained by summing the compressions of all layers.

The shortcomings of layer-wise summation method can be analyzed from the following factors: the calculation and distribution of additional stress, the selection of compression indicators, and the thickness of the soil layers. The method adopts the elastic theory to calculate the vertical stress σ_z and uses e-p curve to determine the uniaxial compression and deformation, which is deviated from the stress condition of ground base.

For the deformation index, the test condition determines its results, and the selection method also affects the calculation results. Furthermore, for the thickness of compression layer, there is no strict theoretical basis. The determination of the thickness is based on a semi-empirical method, which is only verified from engineering measurement. The above factors results in the fact that the different methods that determine the thickness of compressed layer can produce the error of 10%. However, the concept of the layer-wise summation method of settlement calculation is clear. The calculation process and deformation index is relatively simple and easy to be grasped. So it is still widely used in engineering.

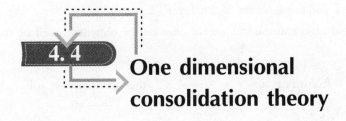

4.4 One dimensional consolidation theory

The mechanics of the one-dimensional consolidation process can be represented by means of a simple analogy as shown in Fig. 4.4.1.

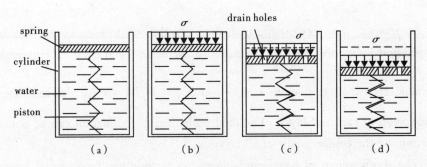

Fig. 4.4.1 Consolidation analogy

Figure 4.4.1a shows a spring inside a cylinder filled with water and a piston, on top of the spring. It is assumed that there can be no leakage between the piston and the cylinder and no friction. The spring represents the compressible soil

Chapter 4 Compression and consolidation of soils

skeleton, the water in the cylinder the pore water and the bore diameter of the valve the permeability of the soil. The cylinder itself simulates the condition of no lateral strain in the soil. Suppose a load is now placed on the piston with no leakage, as in Fig. 4.4.1b. Assuming water to be incompressible, the piston will not move, with the result that no load can be transmitted to the spring; the load will be carried by the water, the increase in pressure in the water being equal to the load divided by the piston area. This situation corresponds to the undrained condition in the soil. If the drain holes are now opened, water will be forced out through the drain holes at a rate governed by the bore diameter. This will allow the piston to move and the spring to be compressed as load is gradually transferred to it. This situation is shown in Fig. 4.4.1c. At any time the increase in load on the spring will correspond to the reduction in pressure in the water. Eventually, as shown in Fig. 4.4.1d, all the load will be carried by the spring and the piston will come to rest, this corresponding to the drained condition in the soil. At any time, the load carried by the spring represents the effective normal stress in the soil, the pressure of the water in the cylinder the pore water pressure and the load on the piston the total normal stress. The movement of the piston represents the change in volume of the soil and is governed by the compressibility of the spring (the equivalent of the compressibility of the soil skeleton). The piston and spring analogy represents only an element of soil since the stress conditions vary from point to point throughout a soil mass.

(1) Terzaghi's theory of one-dimensional consolidation

The assumptions made in the theory of one-dimensional consolidation are:

① The soil is homogeneous and fully saturated.

② The solid particles and water are incompressible.

③ Compression and flow are one-dimensional (vertical).

④ Strains are small.

⑤ Darcy's law is valid at all hydraulic gradients.

⑥ The coefficient of permeability and the coefficient of volume compressibility remain constant throughout the process.

⑦ There is a unique relationship, independent of time, between void ratio and effective stress.

Consider an element of fully saturated soil having dimensions dx, dy and dz in the x, y and z directions, respectively, within a clay layer of thickness H, as shown in Fig. 4.4.2. An increment of total vertical stress p_0 is applied to the element with flow taking place in the z direction only. The component of discharge velocity of water entering the element is v_z, and the rate of change of discharge velocity in the z direction is $\partial v_z / \partial z$. The volume of water entering the element per unit time is

$$Q_e = v_z dx dy \qquad (4.4.1)$$

And the volume of water leaving per unit time is

$$Q_l = v_z dx dy + \frac{\partial v_z dx dy}{\partial z} dz \qquad (4.4.2)$$

If water is assumed to be incompressible, the difference between the volume of water entering the element per unit time and the volume leaving is

$$\Delta Q = Q_l - Q_e = \frac{\partial v_z}{\partial z} dx dy dz \qquad (4.4.3)$$

The flow velocity through the element is given by Darcy's law as

$$v_z = ki = -k \frac{\partial h}{\partial z} \qquad (4.4.4)$$

Since any change in total head (h) is due only to a change in pore water pressure

$$u = \gamma_w h, \quad h = \frac{u}{\gamma_w} \qquad (4.4.5)$$

$$v_z = ki = -k \frac{\partial h}{\partial z} = -\frac{k}{\gamma_w} \frac{\partial u}{\partial z} \qquad (4.4.6)$$

$$\frac{\partial v_z}{\partial z} = -\frac{k}{\gamma_w} \frac{\partial^2 u}{\partial z^2} \qquad (4.4.7)$$

$$\Delta Q = \frac{\partial v_z}{\partial z} dx dy dz = -\frac{k}{\gamma_w} \frac{\partial^2 u}{\partial z^2} dx dy dz \qquad (4.4.8)$$

According to the condition of continuity, the equation of continuity can therefore be expressed as

$$\Delta Q = -\frac{k}{\gamma_w} \frac{\partial^2 u}{\partial z^2} dx dy dz = \frac{dV_v}{dt} \qquad (4.4.9)$$

where dV_v / dt is the volume change per unit time.

The rate of volume change can be expressed as

Chapter 4 Compression and consolidation of soils

$$\frac{dV_v}{dt} = \frac{\partial}{\partial t}\left(\frac{e}{1+e_0}dxdydz\right) = \frac{1}{1+e_0}\frac{\partial e}{\partial t}dxdydz = m_v\frac{\partial \sigma'}{\partial t}dxdydz \quad (4.4.10)$$

The total stress increment is gradually transferred to the soil skeleton, increasing effective stress, as the excess pore water pressure decreases. Hence the rate of volume change can be expressed as

$$\partial \sigma' = \partial(\sigma - u) = -\partial u \quad (4.4.11)$$

$$\frac{dV_v}{dt} = m_v\frac{\partial u}{\partial t}dxdydz \quad (4.4.12)$$

Combining Equations (4.4.9) and (4.4.12),

$$\frac{k}{\gamma_w}\frac{\partial^2 u}{\partial z^2}dxdydz = m_v\frac{u}{t}dxdydz \quad (4.4.13)$$

or

$$\frac{\partial u}{\partial t} = \frac{k}{m_v\gamma_w}\frac{\partial^2 u}{\partial z^2} = C_v\frac{\partial^2 u}{\partial z^2} \quad (4.4.14)$$

This is the differential equation of consolidation, in which C_v being defined as the coefficient of consolidation, suitable unit being m^2/year. Since k and m_v are assumed as constants, C_v is constant during consolidation.

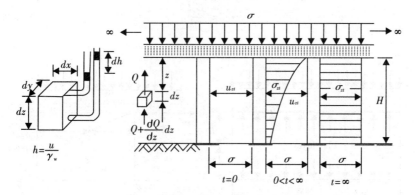

Fig. 4.4.2 Seepage through a soil element

(2) Solution of the consolidation equation

Consider the initial condition and boundary conditions of a clay layer of thickness $2H$, as shown in Fig. 4.4.3, the solution for the differential equation of consolidation is as follows.

① The total stress increment is assumed to be applied instantaneously. At zero time, therefore, the increment will be carried entirely by the pore water, i.e. the initial value of excess pore water pressure (u) is equal to σ and the initial condition is

$$u = \sigma, \sigma' = 0 \text{ for } 0 \leq z \leq 2H \text{ when } t = 0 \tag{4.4.15}$$

② The upper and lower boundaries of the clay layer are assumed to be free-draining, the permeability of the soil adjacent to each boundary being very high compared to that of the clay. Thus the boundary conditions at any time after the application of σ are

$$u = 0, \sigma' = \sigma \text{ for } z = 0 \text{ and } z = 2H \text{ when } 0 < t < \infty \tag{4.4.16}$$

$$u = 0, \sigma' = \sigma \text{ for } 0 \leq z \leq 2H \text{ when } t = \infty \tag{4.4.17}$$

The solution for the excess pore water pressure at depth z after time t is

$$u_{zt} = \frac{4\sigma}{\pi} \sum_{m=1}^{\infty} \frac{1}{m} e^{-\frac{m^2\pi^2}{4} T_v} \sin \frac{m\pi z}{2H} \tag{4.4.18}$$

or

$$u_{zt} = 2\sigma \sum_{m=1}^{\infty} \frac{1}{M} e^{-M^2 T_v} \sin \frac{Mz}{H}, \quad M = \frac{m\pi}{2} \tag{4.4.19}$$

Where m are odd values. T_v is a dimensionless number called the time factor.

$$T_v = \frac{C_v t}{H^2} \tag{4.4.20}$$

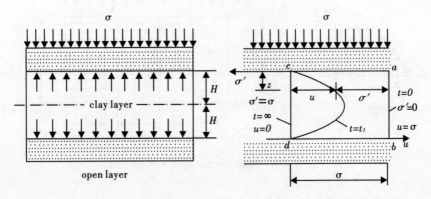

Fig. 4.4.3 Distribution of excess pore water pressure

The progress of consolidation can be shown by plotting a series of curves of u against z for different values of t. Such curves are called isochrones and their form will depend on the initial distribution of excess pore water pressure and the drainage conditions at the boundaries of the clay layer.

The degree of consolidation can be expressed as

$$U_{zt} = \frac{\sigma'_{zt}}{\sigma} = \frac{\sigma - u_{zt}}{\sigma} = 1 - \frac{u_{zt}}{\sigma} \tag{4.4.21}$$

Combining eqs. (4.4.18) and (4.4.21), or Combining Eqs. (4.4.19) and (4.4.21)

$$U_{zt} = 1 - \frac{u_{zt}}{\sigma} = 1 - \frac{4}{\pi} \sum_{m=1}^{\infty} \frac{1}{m} e^{-\frac{m^2\pi^2}{4}T_v} \sin\frac{m\pi z}{2H} \tag{4.4.22}$$

or

$$U_{zt} = 1 - \frac{u_{zt}}{\sigma} = 1 - \sum_{m=1}^{\infty} \frac{2}{M} e^{-M^2 T_v} \sin\frac{Mz}{H}, \quad M = \frac{m\pi}{2} \tag{4.4.23}$$

In practical problems it is the average degree of consolidation (U_t) over the depth of the layer as a whole that is of interest, the consolidation settlement at time t (s_t) being given by the product of U_t and the final settlement (s). The average degree of consolidation at time t for constant σ is given by

$$U_t = \frac{s_t}{s} = \frac{m_v \int_0^{2H} \sigma' dz}{m_v \int_0^{2H} \sigma dz} = \frac{\int_0^{2H} (\sigma - u_{zt}) dz}{\int_0^{2H} \sigma dz} = 1 - \frac{\int_0^{2H} u_{zt} dz}{2H\sigma} \tag{4.4.24}$$

Combining Eqs. (4.4.19) and (4.4.24)

$$U_t = 1 - \sum_{m=1}^{\infty} \frac{2}{M^2} e^{-M^2 T_v} \tag{4.4.25}$$

or

$$U_t = 1 - \frac{8}{\pi^2} e^{-M^2 T_v} = 1 - \frac{8}{\pi^2} e^{-\frac{\pi^2}{4}T_v} \text{ when } m = 1 \tag{4.4.26}$$

The relationship between U_t and T_v is also represented by curves, as shown in Fig. 4.4.4, where the parameter α equals σ'_z/σ''_z, σ'_z is the consolidation pressure on the pervious boundary surface and σ''_z is the consolidation pressure on the impervious boundary surface. For the case of two way drainage, i.e., open layer, the average degree of consolidation can also be calculated using Fig. 4.4.4, but α is equals to 1 and the drainage path H should be half the thickness of the layer.

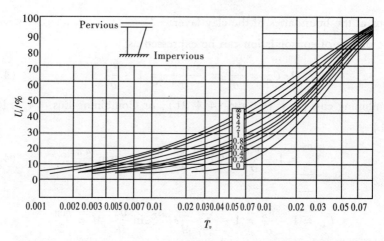

Fig. 4.4.4 Relationship of overage degree of consolidation and time factor

According to Eq. (4.4.26), with the distribution of consolidation pressure and the drainage condition, the following two types of engineering problem can be resolved.

(1) With known final settlement s, find settlement s_t at an elapsed consolidation time t.

For such problem, firstly the average coefficient of consolidation C_v, and the time factor T_v of the soil layer can be calculated using the known values of k, a, e_1, H and t. Then the corresponding average degree of consolidation U_t can be read off from Fig. 4.4.4. Therefore, s_t can be calculated.

(2) With known final settlement s, find the length of time t for achieving a settlement s_t.

For such problem, firstly the average degree of consolidation U_t is s_t/s, then the time factor T_v of the soil layer can be read off from Fig. 4.4.4. Therefore, the length of time t required can be calculated using Eq. (4.4.26).

EXERCISES

4.1 When the compression pressure is 100kPa and 200kPa, the void ratio is 0.944 and 0.935 obtained from the oedometer test results. Find the compression coefficient a_{1-2} and the compression modulus and judge its compressibility.

Chapter 4 Compression and consolidation of soils

4.2 The thickness of the clay layer is 4m and its initial void ratio is 1.25. If the infinity uniform load 100kPa is applied on its surface, the void ratio equals to 1.12 when the settlement is stable. Try to find the settlement of the clay layer.

4.3 There is a clay layer located between two sand layers. The thickness of the clay is 4m. A specimen of 2cm thickness is taken from the center of the clay of the oedometer test with porous stones on both surfaces of the specimen. It is measured that 7min is required to achieve a degree of consolidation of 80%. The infinity uniform load 100kPa is instantly applied on the surface of the natural clay layer. How long it will take for the natural clay layer to achieve a degree of consolidation of 80%.

4.4 A clay specimen is taken from a position below the ground surface, where the soil is subjected to an effective stress of 100kPa. The initial void ratio of the specimen is 1.05. From the oedometer test, the relationship between the effective stress and the void after consolidation is listed as follows (Table 4.1). Plot the compression, swelling and recompression curves in a semi lg scale. Derive the preconsolidation pressure p_c and the values of C_s and C_c of in-situ compression curve. Determine the types of consolidation of such soil.

Table 4.1 Exercise 4.4

Loading	p(kPa)	50	100	200	400
	e	0.950	0.922	0.888	0.835
Unloading	p(kPa)	200	100		
	e	0.840	0.856		
Reloading	p(kPa)	200	400	800	1600
	e	0.845	0.830	0.757	0.675

4.5 The width of a rectangular foundation is 4m. The additional stress on the floor is 100kPa. Its embedment depth is 2m. There is two soil layers in the depth scope 12m. The thickness of the upper soil is 6m and its bulk unit weight is 18kN/m³ and the relationship between its void ratio and compression pressure is $e = 0.85 - \frac{2}{3}p$. The thickness of the subsoil is 6m and its bulk unit weight is 20kN/m^3 and the relationship between its void ratio and compression pressure is $e = 1.0 - p$. The water table is 6m. The additional stress coefficient and the average additional stress coefficient distribution under the center of the foundation are shown in table 4-2. Find the foundation settlement used the layer-wise summation method and the recommendation settlement method by code (the experience coefficient of settlement is 1.05).

Table 4.2 The additional and the average additional stress coefficient

Depth/m	0	1	2	3	4	5
The additional stress coefficient	1.0	0.94	0.75	0.54	0.39	0.28
The average additional stress coefficient	1.0	0.98	0.92	0.83	0.73	0.65
Depth/m	6	7	8	9	10	
The additional stress coefficient	0.21	0.17	0.13	0.11	0.09	
The average additional stress coefficient	0.59	0.53	0.48	0.44	0.41	

4.6 The thickness of a clay layer is 4m and it is an half-closed layer. The infinity uniform load is applied on the surface and the final settlement is 28cm. After 100 days, the settlement of the clay layer is 18.5cm and the relationship between the degree of consolidation and time factor is $U = 1.128 (T_v)^{\frac{1}{2}}$. Try to find the coefficient of consolidation C_v.

REFERENCES

1. Soil Mechanics Work Team at Hohai University. Soil Mechanics [M]. China Communication Press, Beijing, 2004.
2. Siwei He. Essentials of Soil Mechanics [M]. Zhongshan University Press, Guangzhou, 2003.
3. Ministry of Construction P. R. China. GB50021 −2001 Code for Investigation of Geotechnical Engineering (2009 revised edition) [S]. Beijing: China Building Industry Press, 2009.
4. Ministry of Construction P. R. China. GB50025 − 2007 Code for Building Construction in Collapsible Loess Zone[S]. Beijing: China Building Industry

Press, 2004.
5. Ministry of Water Resources P. R. China. GB/T50145 −2007 Standard for Engineering Classification of Soil[S]. Beijing: China Planning Press, 2008.
6. P. R. China Ministry of Housing and Urban-Rural Development Producer. GB50007 −2011 Code for Design of Building Foundation[S]. Beijing: China Building Industry Press, 2009.
7. Nanjing Hydraulic Research Institute. SL237 −1999 Specification of Soil Test [S]. Beijing: China Waterpower Press, 1999.
8. Ministry of Water Resources P. R. China. GB/T 50123 −1999 Standard for Soil Test Method[S]. Beijing: China Planning Press, 2000.
9. Shude Zhao, Hongjian Liao. Soil Mechanics (the second edition) [M]. Beijing: Higher Education Press, 2010.
10. Shude Zhao, Hongjian Liao. Civil Engineering Geology[M]. Beijing: Science Press, 2009.

SOIL MECHANICS
CHAPTER 5
Shear strength

5.1 Shear resistance

Soil is a particulate material, so shear failure occurs when stresses between the particles are such that they slide or roll each other. Hence, the shear strength of soil is mainly controlled by friction. If at a point on any plane within a soil mass the shear stress becomes equal to the shear strength of the soil, the failure will occur at that point, as shown in Fig. 5.1.1.

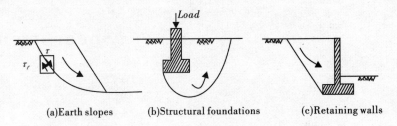

(a) Earth slopes   (b) Structural foundations   (c) Retaining walls

Fig. 5.1.1 Typical applications of strength in soils

The shear resistance between two particles is the force that must be applied to cause a relative movement between the particles. If N is the normal force across a surface, the maximum shear force on this surface is proportional to the normal force, as shown in Fig. 5.1.2(a).

$$T_{max} = N\mu = N\tan\varphi \tag{5.1.1}$$

where μ is the coefficient of friction. φ is a friction angle defined as

$$\tan\varphi = \mu \tag{5.1.2}$$

The resistance of soil to deformation is influenced strongly by the shear resistance at contact between particles. It is generally assumed that the relation between the normal stress σ on every section through a mass of cohesive soil and the corresponding shearing resistance τ per unit of area can be represented by an empirical equation

$$\tau = c + \sigma \tan\varphi \quad (5.1.3)$$

where the symbol c represents the cohesion, which is equal to the shearing resistance per unit of area if $\sigma = 0$. The equation is known as Coulomb's equation. For cohesionless soil ($c = 0$) as shown in Fig. 5.1.2(b), the corresponding equation is

$$\tau = \sigma \tan\varphi \quad (5.1.4)$$

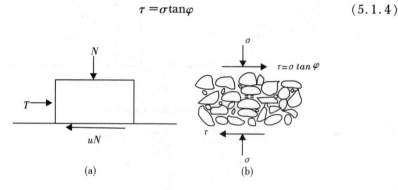

Fig. 5.1.2 Shear resistance

The values c and φ contained in the preceding equations can be determined by laboratory tests, by measuring the shearing resistance on plane sections through the soil at different values of the normal stress σ. The states of stress in two dimensions can be represented on a plot of shear stress (τ) against normal stress (σ). The relation between σ and τ is shown in Fig. 5.1.3. The shear strength (τ_f) of a soil at a point on a particular plane is also expressed by Coulomb as a linear function of the normal stress at failure (σ_f) on the plane at the same point.

$$\tau_f = c + \sigma_f \tan\varphi \quad \text{for cohesive soil} \quad (5.1.5)$$

$$\tau_f = \sigma_f \tan\varphi \quad \text{for cohesionless soil} \quad (5.1.6)$$

The c and φ are also called shear strength parameters referred to as the cohesion intercept and the angle of shearing resistance, respectively.

However, in accordance with the principle of effective stress, the shear stress in a soil can be resisted only by the skeleton of solid particles. The shear strength should be expressed as a function of effective normal stress at failure (σ_f'), and the effective shear strength parameters are denoted c' and φ' and need to be determined by either laboratory or in-situ tests.

$$\tau_f = c' + \sigma_f' \tan\varphi' \quad \text{for cohesive soil} \quad (5.1.7)$$

$$\tau_f = \sigma'_f \tan\varphi' \quad \text{for cohesionless soil} \tag{5.1.8}$$

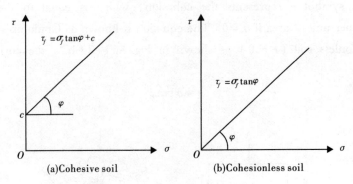

Fig.5.1.3 Relation between σ and τ

5.2 Mohr-Coulomb failure criterion

A stress state can be represented either by a point with coordinates τ and σ (σ') as shown in Fig.5.2.1, or by a Mohr circle defined by the effective principal stresses $\sigma_1(\sigma_1')$ and $\sigma_3(\sigma_3')$. Mohr circles representing stress states at failure are shown in Fig.5.2.2, compressive stress being taken as positive. The line touching the Mohr circles may be straight or slightly curved and is referred to as the failure envelope. A state of stress represented by a Mohr circle part of which lies above the envelope, is impossible. For a given state of stress it is apparent that, because $\sigma_1' = \sigma_1 - u$ and $\sigma_3' = \sigma_3 - u$, the Mohr circles for total and effective stresses have the same diameter but their centres are separated by the corresponding pore water pressure u. Similarly, the total and effective stress points are separated by the value of u.

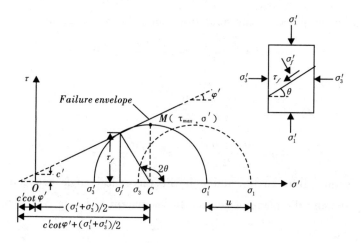

Fig. 5.2.1　Mohr-Coulomb failure criterion

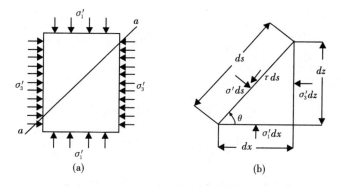

Fig. 5.2.2　Stress conditions in soil during triaxial compression test

In order to determine the stresses on an arbitrary inclined section a-a through the specimen shown in Fig. 5.2.2(a), we investigate the conditions for the equilibrium of a small prism shown in Fig. 5.2.2(b), one side of which is located on the inclined section. The other two sides are parallel to the direction of the principal stresses σ_1' and σ_3'. The slope of the inclined surface is determined by the angle θ. The equilibrium of the prism requires that

$$\sum F_x = \sigma' \sin\theta ds - \sigma_3' \sin\theta ds - \tau\cos\theta ds = 0 \qquad (5.2.1)$$

$$\sum F_z = \sigma_1' \cos\theta ds - \sigma' \cos\theta ds - \tau\sin\theta ds = 0 \qquad (5.2.2)$$

Solving above equations, the values of the stresses σ and τ on the slope surface can be computed by the numerical values for σ_1', σ_3' and θ:

$$\sigma' = \frac{\sigma_1' + \sigma_3'}{2} + \frac{\sigma_1' - \sigma_3'}{2}\cos 2\theta \qquad (5.2.3)$$

$$\tau = \frac{\sigma_1' - \sigma_3'}{2}\sin 2\theta \qquad (5.2.4)$$

or

$$(\sigma' - \frac{\sigma_1' + \sigma_3'}{2})^2 + \tau^2 = (\frac{\sigma_1' - \sigma_3'}{2})^2 \qquad (5.2.5)$$

This is the Mohr circle, and θ is the theoretical angle between the major principal plane and the plane of failure. It is apparent as in Fig. 5.2.1 that

$$2\theta = 90° + \varphi' \qquad (5.2.6)$$

$$\theta = 45° + \frac{\varphi'}{2} \qquad (5.2.7)$$

$$\sin\varphi' = \frac{\frac{1}{2}(\sigma_1' - \sigma_3')}{c'\cot\varphi' + \frac{\sigma_1' + \sigma_3'}{2}} = \frac{\sigma_1' - \sigma_3'}{2c'\cot\varphi' + \sigma_1' + \sigma_3'} \qquad (5.2.8)$$

Therefore

$$\sigma_1' = \sigma_3'\tan^2(45° + \frac{\varphi'}{2}) + 2c'\tan(45° + \frac{\varphi'}{2}) \qquad (5.2.9)$$
$$= \sigma_3'\frac{1 + \sin\varphi'}{1 - \sin\varphi'} + 2c'\frac{\cos\varphi'}{1 - \sin\varphi'}$$

$$\sigma_3' = \sigma_1'\tan^2(45° - \frac{\varphi'}{2}) - 2c'\tan(45° - \frac{\varphi'}{2}) \qquad (5.2.10)$$
$$= \sigma_1'\frac{1 - \sin\varphi'}{1 + \sin\varphi'} - 2c'\frac{\cos\varphi'}{1 + \sin\varphi'}$$

These are referred to as the Mohr-Coulomb failure criterion.

The state of stress represented in Fig. 5.2.1 could also be defined by the coordinates of point M, rather than by the Mohr circle. The coordinates of M are $(\sigma_1' - \sigma_3')/2$ and $(\sigma_1' + \sigma_3')/2$, also denoted by τ_{max} and σ', respectively, being the maximum shear stress and the average principal stress. The stress state could also be expressed in terms of total stress. It should be noted that

$$\frac{\sigma_1' - \sigma_3'}{2} = \frac{\sigma_1 - \sigma_3}{2} \qquad (5.2.11)$$

Chapter 5 Shear strength

$$\frac{\sigma_1' + \sigma_3'}{2} = \frac{\sigma_1 + \sigma_3}{2} - u \qquad (5.2.12)$$

Stress point M lies on a modified failure envelope as shown in Fig. 5.2.3 defined by the equation

$$q = a' + p' \tan\alpha' \qquad (5.2.13)$$

where a' and α' are the modified shear strength parameters, the relation between parameters c' and φ' are given approximately by Fig. 5.2.3.

$$\tan\alpha' = \sin\varphi' = \frac{R}{O'A}, \ \alpha' = \tan^{-1}(\sin\varphi') \qquad (5.2.14)$$

$$\tan\varphi' = \frac{c'}{O'O} \qquad (5.2.15)$$

$$\tan\alpha' = \frac{a'}{O'O} = \frac{a'}{c'}\tan\varphi' = \frac{a'\sin\varphi'}{c'\cos\varphi'} \qquad (5.2.16)$$

Combining equations (5.2.14) and (5.2.16)

$$a' = c'\cos\varphi' \qquad (5.2.17)$$

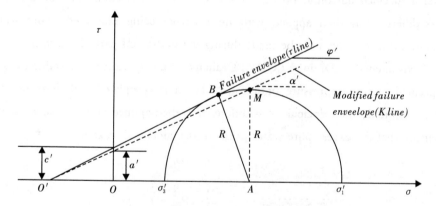

Fig. 5.2.3 Relation between τ line and K line

5.3 Shear strength tests

Many variations of shear strength test procedure are possible but the three principal types of test are listed as follows:

Unconsolidated - Drained (UU). The specimen is subjected to a specified all-round pressure and then the principal stress difference is applied immediately, with no drainage being permitted at any stage of the test.

Consolidated - Undrained (CU). Drainage of the specimen is permitted under a specified all-round pressure until consolidation is complete; the principal stress difference is then applied with no drainage being permitted. Pore water pressure measurements may be made during the undrained part of the test.

Consolidated - Drained (CD). Drainage of the specimen is permitted under a specified all-round pressure until consolidation is complete; with drainage still being permitted, the principal stress difference is then applied at a rate slow enough to ensure that the excess pore water pressure is maintained at zero.

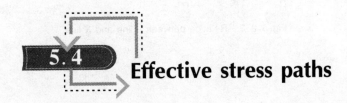

5.4 Effective stress paths

The successive states of stress in a test specimen can be represented by a series of Mohr circles or, by a series of stress points, as shown in Fig. 5.4.1. The curve or straight line connecting the relevant stress points is called the stress paths, gives a clear representation of the successive states of stress. Stress paths may be drawn in terms of either effective or total stresses. The horizontal distance between the effective and total stress paths is the value of pore water pressure. The effective and total stress paths (denoted by ESP and TSP, respectively) for the triaxial tests are shown in Fig. 5.4.1, the coordinates being $(\sigma_1 - \sigma_3)/2$ and $(\sigma_1' + \sigma_3')/2$ or the total stress equivalents.

$$q = \frac{\sigma_1' - \sigma_3'}{2} \tag{5.4.1}$$

$$p' = \frac{\sigma_1' + \sigma_3'}{2} \tag{5.4.2}$$

The effective stress paths terminate on the modified failure envelope. The effective stress paths for the drained tests and all the total stress paths are straight lines at a slope of 45°. The detailed shape of the effective stress paths for the consolidated-undrained tests depends on the pore water pressure u.

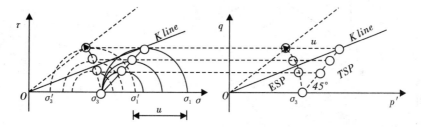

Fig. 5.4.1 Stress path

Stress paths are also plotted with respect to deviator stress $(\sigma_1 - \sigma_3)$ and average effective principal stress $(\sigma_1' + \sigma_2' + \sigma_3')/3$, denoted by q and p', respectively. In the triaxial test the intermediate principal stress (σ_2') is equal to the minor principal stress (σ_3'), therefore

$$q = \sigma_1' - \sigma_3' = \sigma_1 - \sigma_3 \tag{5.4.3}$$

$$p' = \frac{\sigma_1' + 2\sigma_3'}{3} = \sigma_3 + \frac{1}{3}q - u \tag{5.4.4}$$

The effective and total stress paths for the triaxial tests are shown in Fig. 5.4.2, the coordinates being $(\sigma_1 - \sigma_3)$ and $(\sigma_1' + 2\sigma_3')/3$ or the total stress equivalents.

The equation of the projection of the critical state line in Fig. 5.4.2 on the q-p' plane is

$$q = Mp' \tag{5.4.5}$$

where M is the slope of C. S. L., the parameter M can be related to the angle of shearing resistance φ'

$$M = \frac{6\sin\varphi'}{3 - \sin\varphi'} \tag{5.4.6}$$

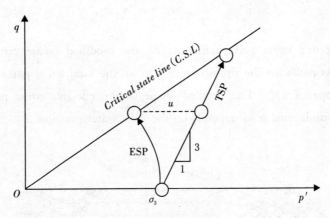

Fig. 5.4.2 Stress path

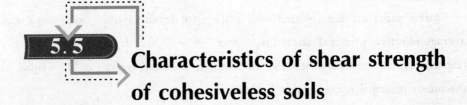

Characteristics of shear strength of cohesiveless soils

The main factor influencing the shear strength of sand is the initial compactness. The initial compactness could be reflected by the size of void ratio. In general, the smaller the initial void ratio and the closer contact of particles, the greater sliding and rolling friction between the particles is and the greater friction resistance is. Thus the shear strength is greater.

Figure 5.5.1 shows the stress-strain relationship and the volume change of the sand under the same confining pressure σ_3 with different initial void ratio during shearing. Obviously, the void ratio of compacted dense sand is smaller, and there is an obvious peak on stress-strain relationship curve. The stress begins to decrease with the increase of the strain after the peak. This feature is called the strain softening. The strength at the peak is called peak strength, and the strength at the final stable value is called the residual strength. When the dense sand shearing, its volume decreases slightly in the beginning, and then increases significantly (dilatancy). Generally, the volume is greater than the initial volume. This feature is called the dilatancy of soil. The strength of the loose sand increases with the increase of axial strain, and generally there is no peak on the curve. This feature is called the strain hardening. When the loose sand shearing, the particles scroll to the equilibrium position and arrange more closely, and then the volume decreases. This feature is called the shear shrinkage. With the increase of pressure and the crush of soil particles, the dilatancy trend of dense sand will gradually disappear. So under the high confining pressure, no matter how elastic the sand is, the shear shrinkage will occur.

For shear tests of dense sands there is a tendency of the specimen to dilate as the test progresses. Similarly, in loose sand the volume gradually decreases. An increase or decrease of volume means a change in the void ratio of soil. The nature of the change of the void ratio with strain for loose and dense sands is shown in Fig. 5.5.2. The void ratio for which the change of volume remains constant during shearing is called the critical void ratio. Figure 5.5.3 shows the results of some drained triaxial tests on washed Fort Peck sand. The void ratio after the application of σ_3 is plotted in the ordinate, and the change of volume V at the peak point of the stress-strain plot, is plotted along the abscissa. For a given σ_3, the void ratio corresponding to $V=0$ is the critical void ratio. Note that the critical void ratio is a function of the confining pressure σ_3. It is, however, necessary to recognize that, whether the volume of the soil specimen is increasing or decreasing, the critical void ratio is reached only in the shearing zone, even if it is generally calculated on the basis of the total volume change of the specimen. The concept of critical void ratio was first introduced in 1938 by Casagrande to study liquefaction of granular soils. When a natural deposit of saturated sand that has a void ratio greater than the critical void ratio is subjected to a sudden shearing stress (due to an earthquake or to blasting, for example), the sand will undergo a decrease in volume. This will result in an increase of pore water pressure u. At a given depth, the effective stress is given by the relation $\sigma' = \sigma - u$. If σ (i.e., the total stress) remains constant and u increases, the result will be a decrease in σ'. This, in turn, will reduce the shear strength of the soil. If the shear strength is reduced to a value which is less than the applied shear stress, the soil will fail. This is called soil liquefaction.

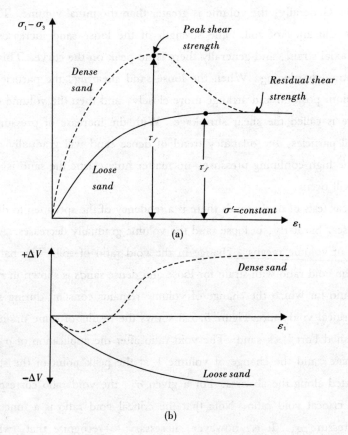

Fig. 5.5.1 Experimental results of sand

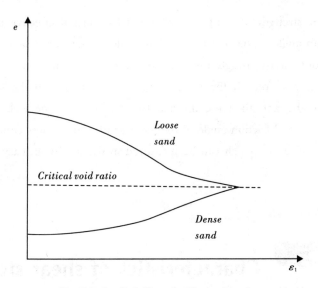

Fig. 5.5.2 Definition of critical void ratio

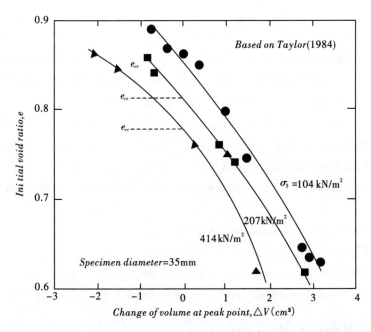

Fig. 5.5.3 Critical void ratio from triaxial test on Fort Peck sand

The shear strength of sand is determined by normal effective stress and the internal friction angle. The internal friction angle of the dense sand is relevant to the initial void ratio, roughness on the surface of the soil particles, grain composition, and so on. If the initial void ratio is small and the surface of soil particles is rough, and then the internal friction angle of the well-graded soil is greater. The internal friction angle of the loose sand more or less equals the natural rest angle of dry sand, which can be measured in the lab by an easy way.

5.6 Characteristics of shear strength of cohesive soils

The shear strength of cohesive soils can generally be determined in the laboratory by either direct shear test equipment or triaxial shear test equipment; however, the triaxial test is more commonly used. Only the shear strength of saturated cohesive soils will be presented here. The shear strength based on the effective stress can be given by $\tau = \sigma' \tan \varphi' + c'$. For normally consolidated clays, $c \approx 0$, and for overconsolidated clays, $c > 0$.

Three conventional types of tests are conducted with cohesive soils in the laboratory:
(1) Consolidated drained test (CD test).
(2) Consolidated undrained test (CU test).
(3) Unconsolidated undrained test (UU test).
Each of these tests will be separately considered in the following sections.

Consolidated drained test

For the consolidated drained test, the saturated soil specimen is first subjected to a confining pressure through the chamber fluid. The connection to the drainage is kept open for complete drainage, so that the pore water pressure is equal to zero. Then the deviator stress is increased at a very slow rate, keeping the drainage valve open to allow complete dissipation of the resulting pore water pressure. Figure 5.6.1 shows the nature of the variation of the deviator stress with axial strain. From Fig. 5.6.1, it must also be pointed out that, during the application of the deviator stress, the volume of the specimen gradually reduces for normally consolidated clays. However, the overconsolidated clays go through some reduction of volume initially but then expand. In a consolidated drained test the total stress is equal to the effective stress, since the excess pore water pressure is zero.

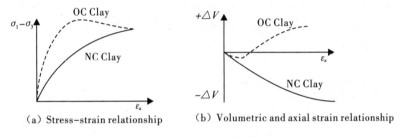

(a) Stress-strain relationship (b) Volumetric and axial strain relationship

Fig. 5.6.1 Consolidated drained triaxial test in clay

From the results of a number of tests conducted using several specimens, Mohr's circles at failure can be plotted as shown in Fig. 5.6.2. The values of c and φ are obtained by drawing a common tangent to Mohr's circles, which is the Mohr-Coulomb envelope. For normally consolidated clays, $c = 0$. Thus the equation of the Mohr-Coulomb envelope can be given by $\tau_f = \sigma' \tan\varphi$. The slope of the failure envelope will give us the angle of friction of the soil. As shown by Eq. (5.6.1):

$$\sin \varphi = \left(\frac{\sigma_1' - \sigma_3'}{\sigma_1' + \sigma_3'}\right)_{failure} \quad (5.6.1)$$

For the overconsolidated clays (Fig. 5.6.2b), $c \neq 0$. So the shear strength follows the equation $\tau_f = \sigma' \tan\varphi + c$. The values of c and φ can be determined by measuring the intercept of the failure envelope on the shear stress axis and the slope of the failure envelope, respectively. The formula is obtained:

$$\sin\varphi = \frac{\sigma_1' - \sigma_3'}{(\sigma_1' + \sigma_3') + c\cot\varphi} \quad (5.6.2)$$

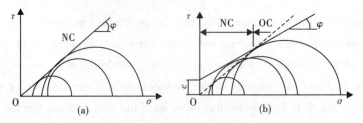

Fig.5.6.2 Failure envelope for drained test

Figure 5.6.1 shows that at very large strains the deviator stress reaches a constant value. The shear strength of clays at very large strains is referred to as residual shear strength (i.e., the ultimate shear strength). It has been proved that the residual strength of a given soil is independent of past stress history. The residual friction angle in clays is of importance in subjects such as the long-term stability of slopes.

Consolidated undrained test

In the consolidated undrained test, the soil specimen is first consolidated by a chamber-confining pressure; full drainage from the specimen is allowed. After complete dissipation of excess pore water pressure generated by the confining pressure, the deviator stress is increased to cause failure of the specimen. During this phase of loading, the drainage line from the specimen is closed. Since drainage is not permitted, the pore water pressure in the specimen increases. Figure 5.6.3 shows the results of the consolidated undrained test. The pore water pressure at failure u_f is positive for normally consolidated clays and becomes negative for overconsolidated clays. Thus u_f is dependent on the overconsolidation ratio. The overconsolidation ratio, OCR, for triaxial test conditions may be defined as

$$OCR = \frac{\sigma'_3}{\sigma_3} \qquad (5.6.3)$$

where $\sigma'_3 = \sigma_3$ is the maximum chamber pressure at which the specimen is consolidated and then allowed to rebound under a chamber pressure of σ_3.

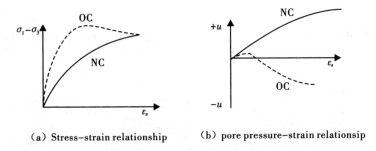

(a) Stress–strain relationship   (b) pore pressure–strain relationsip

Fig. 5.6.3 Results of consolidated undrained test

Consolidated undrained tests on a number of specimens can be conducted to determine the shear strength parameters of a soil, for the case of normally consolidated clay, as shown in Fig. 5.6.4. The total-stress Mohr's circle is shown by solid line. The effective stress Mohr's circle is shown by dashed line. A common tangent drawn to the effective-stress circle will give the Mohr-Coulomb failure envelope given by the equation $\tau = \sigma' \tan\varphi$. If we draw a common tangent to the total-stress circles, it will be a straight line passing through the origin. This is the total-stress failure envelope, and it may be given by

$$\tau = \sigma \tan\varphi_{cu} \qquad (5.6.4)$$

where φ_{cu} is the consolidated undrained angle of friction.

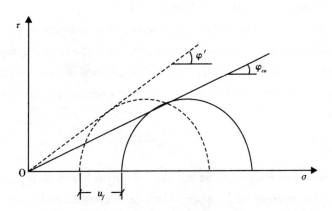

Fig. 5.6.4 Failure envelop for normally consolidated clay

The total-stress failure envelope for an overconsolidated clay will be of the nature shown in Fig. 5.6.5 and can be given by the relation

$$\tau = \sigma \tan\varphi_{cu} + c_{cu} \quad (5.6.5)$$

where c_{cu} is the intercept of the total-stress failure envelope along the shear stress axis.

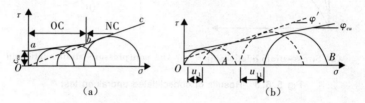

Fig. 5.6.5 Failure envelop for overconsolidated clay

The shear strength parameters for overconsolidated clay based on effective stress, i.e., c' and φ', can be obtained by plotting the effective-stress Mohr's circle and then drawing a common tangent.

$$\tau = \sigma' \tan\varphi' + c' \quad (5.6.6)$$

Unconsolidated undrained test

In unconsolidated undrained triaxial tests, drainage from the specimen is not allowed at any stage. First, the chamber-confining pressure is applied, after which the deviator stress is increased until failure occurs.

Tests of this type can be performed quickly, since drainage is not allowed. For a saturated soil the deviator stress failure is practically the same, irrespective of the confining pressure (Fig. 5.6.6). So the total-stress failure envelope can be assumed to be a horizontal line, and $\varphi = 0$. The undrained shear strength can be expressed as

$$\tau = c_u = \frac{\Delta\sigma_f}{2} \quad (5.6.7)$$

This is generally referred to as the shear strength based on the $\varphi = 0$ concept.

The fact the strength of saturated clays in unconsolidated undrained loading conditions is the same, irrespective of the confining pressure can by explained as following. For saturated soil, $B = 1$ under undrained conditions, the pore water increase $\Delta\sigma_f$ as subjected to an additional confining pressure $\Delta\sigma_f$, while the effective stress keep constant. For a series of experimental test, there is only one effective stress circle, so only the undrained shear strength is measured.

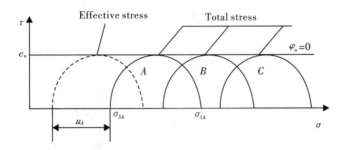

Fig. 5.6.6 Failure envelop of unconsolidated undrained triaxial tests

According to the above analysis, it can be found that, for the total stresses, the strength difference due to the difference of the experiment method can be reflected by strength parameters, i.e., the total stress strength parameters contain the effect of the pore water pressure. The total friction angle satisfy: $\varphi_d > \varphi_{cu} > \varphi_u$. For the effective stresses, the effective strength parameters are almost same.

EXERCISES

5.1 The strength parameters of soil is $c = 20\text{kPa}$, $\varphi = 22°$, the normal stress and shear stress acted on a slope plane are $\sigma = 100\text{kPa}$, $\tau = 60.4\text{kPa}$, respectively. Determine the failure or stability of soil along the plane.

5.2 A series of conventional triaxial consolidated drained tests are conducted on sands. The principal stress difference is $\sigma_1 - \sigma_3 = 400\text{kPa}$, and the confining pressure is $\sigma_3 = 100\text{kPa}$. Determine the strength parameters of sand.

REFERENCES

1. Das B M. Advanced Soil Mechanics (the third edition) [M]. Taylor and Francis, 2008.
2. Craig R F. Craig's Soil Mechanics (the seventh edition) [M]. CRC Press, 2004.
3. Terzaghi K, Peck RB, Mesri G. Soil Mechanics in Engineering Practice (the third edition) [M]. John Wiley & Sons, Inc, 1996.

SOIL MECHANICS CHAPTER 6

Bearing capacity

6.1 Introduction

In this chapter, the bearing capacity of soils will be considered. Loads from a structure are transmitted to the soil through a foundation. The limit equilibrium method of analysis is considered. The limit equilibrium method is used to find solutions for a variety of problems including the bearing capacity of foundations, stability of retaining walls, and slopes.

Before considering these bearing capacity equations, the following definition key terms should be identified:

Foundation is a structure that transmits loads to the underlying soils.

Footing is a foundation consisting of a small slab for transmitting the structural load to the underlying soil. Footings can be individual slabs supporting single columns or combined to support two or more columns, or be a long strip of concrete slab supporting a load bearing wall, or a mat.

Embedment depth is the depth below the ground surface where the base of the foundation rests.

Shallow foundation is one in which the ratio of the embedment depth to the minimum plan dimension, which is usually the width, is $d/B = 2:5$.

Ultimate bearing capacity is defined as the least pressure which would cause shear failure of the supporting soil immediately below and adjacent to a foundation.

Allowable bearing capacity or *safe bearing capacity* is the working pressure that would ensure a margin of safety against collapse of the structure from shear failure. The allowable bearing capacity is usually a fraction of the ultimate bearing capacity.

Three distinct modes of failures have been identified during loading and these are illustrated in Fig. 6.1.1; they will be described with reference to a strip footing. In the case of general shear failure, continuous failure surfaces develop between the edges of the footing and the ground surface as shown in Fig. 6.1.1. As the pressure

Chapter 6 Bearing capacity

increases towards the value q_f, the state of plastic equilibrium is reached initially in the soil around the edges of the footing, and then gradually spreads downwards and outwards. Ultimately the state of plastic equilibrium is fully developed throughout the soil above the failure surfaces. Heaving of the ground surface occurs on both sides of the footing although the final slip movement would occur only on one side, accompanied by tilting of the footing. This mode of failure is typical of low compressibility (i.e. dense or stiff soils) and the pressure-settlement curve is of the general form shown in Fig. 6.1.1, the ultimate bearing capacity being well defined.

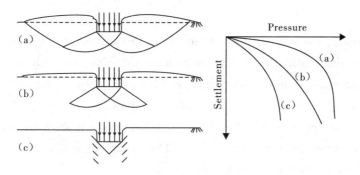

Fig. 6.1.1 Modes of failures
(a) General shear (b) Local shear (c) Punching shear

In the mode of local shear failure, there is significant compression in the soil under the footing and only partial development of the state of plastic equilibrium. The failure surfaces, therefore, do not reach the ground surface and only slight heaving occurs. Tilting of the foundation would not be expected. Local shear failure is associated with high compressibility of soils, as indicated in Fig. 6.1.1, and is characterized by the occurrence of relatively large settlements (which would be unacceptable in practice). The ultimate bearing capacity is not clearly defined.

Punching shear failure occurs when there is compression of the soil under footing, accompanied by shearing in the vertical direction around the edges of the footing. There is no heaving of the ground surface away from the edges and no tilting of the footing. Relatively large settlements are also a characteristic of this mode and the ultimate bearing capacity is not well defined. Punching shear failure will also occur in a soil with low compressibility if the foundation is located at considerable depth. In general the mode of failure depends on the compressibility of soils and the depth of the foundation relative to its breadth.

6.2 Critical edge pressure

Generally, the failure of base begins from the edge of footing. At small load the soil remains elastic. As the load reach a certain value, soils adjacent to footing reach a limit equilibrium condition firstly and the base of the footing will yield. The corresponding load is defined as the critical edge pressure.

A strip footing of width B and embedment depth d is shown in Fig. 6.2.1, which is subjected to a uniform load p. The maximum and minimum principal stresses of an arbitrary element M are:

$$\sigma_{1,3} = \frac{p - \gamma_0 d}{\pi}(\beta_0 \pm \sin \beta_0) \qquad (6.2.1)$$

where β_0 is the angle of M to two endpoints of uniform load (°).

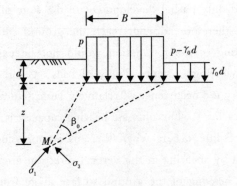

Fig.6.2.1 The principal stresses at uniform load

The gross stress of M is the sum of the gravity stress and surcharge stress. For simplicity, it is assumed that the gravity stress field is regarded as the hydrostatic stress field, i.e. the lateral pressure coefficient equals to 1.0. So the total stress of M is given by

$$\sigma_{1,3} = \frac{p - \gamma_0 d}{\pi}(\beta_0 \pm \sin\beta_0) + \gamma_0 d + \gamma z \qquad (6.2.2)$$

where γ_0 is the average weight above the base, (kN/m^3);

γ is the average weight below the base, (kN/m^3).

When the element M reaches the limit equilibrium condition, according to the Mohr-Coulomb criterion, the following principal stresses exist

$$\frac{\sigma_1 - \sigma_3}{\sigma_1 + \sigma_3 + 2c \cdot \cot\varphi} = \sin\varphi \qquad (6.2.3)$$

Substituting Eq. (6.2.2) into the above equation, the depth of plastic zone is obtained

$$z = \frac{p - \gamma_0 d}{\gamma\pi}\left(\frac{\sin\beta_0}{\sin\varphi} - \beta_0\right) - \frac{c}{\gamma}\cot\varphi - d\frac{\gamma_0}{\gamma} \qquad (6.2.4)$$

where c is the cohesion of soil below base, (kPa);

φ is the friction angle of soil below base, (°).

Equation (6.2.4) is the boundary equation of plastic zone, which reflects the relationship between z and β_0 of the boundary of plastic zone. If d, p, φ, γ and c are known, the plastic zone boundary can be drawn.

The maximum depth of the plastic zone can be obtained by differentiating Eq. (6.2.4)

$$\beta_0 = \frac{\pi}{2} - \varphi \qquad (6.2.5)$$

Substituting Eq (6.2.5) into Eq. (6.2.4), the maximum depth is derived

$$z_{max} = \frac{p - \gamma_0 d}{\gamma\pi}\left(\cot\varphi + \varphi - \frac{\pi}{2}\right) - \frac{c}{\gamma}\cot\varphi - d\frac{\gamma_0}{\gamma} \qquad (6.2.6)$$

According to the above equation, the critical edge pressure is

$$p_{cr} = \gamma_0 d N_q + c N_c \qquad (6.2.7)$$

where

$$N_q = 1 + \frac{\pi}{\cot\varphi + \varphi - \frac{\pi}{2}} \qquad (6.2.8a)$$

$$N_c = \frac{\pi \cot \varphi}{\cot\varphi + \varphi - \frac{\pi}{2}} \qquad (6.2.8b)$$

It can be found that p_{cr} is dependent on φ, γ_0, c and d, and independent on the width of footing.

A large number of engineering practices have shown that p_{cr}, as the bearing capacity design value, is conservative. The experiences have shown that in most cases, even if there is a certain range of plastic zones which do not exceed a permissible range, the structures can be kept in normal use. It is generally believed that the maximum depth of the plastic zone can reach $1/4$ of the width of the footing under a centric loads. Under a small eccentric load, the maximum depth of the plastic zone is allowed to reach $1/3$ of the width of footing. According to the conditions, the corresponding pressures ($p_{1/4}$ and $p_{1/3}$) can be obtained by substituting $z_{max} = 1/4B$ and $z_{max} = 1/3B$ into Eq. (6.2.6), respectively

$$p_{1/4} = \frac{1}{2}\gamma B N_{1/4} + \gamma_0 d N_q + c N_c \qquad (6.2.9a)$$

$$p_{1/3} = \frac{1}{2}\gamma B N_{1/3} + \gamma_0 d N_q + c N_c \qquad (6.2.9b)$$

where

$$N_{1/3} = \frac{\pi}{3\left(\cot\varphi + \varphi - \frac{\pi}{2}\right)} \qquad (6.2.10a)$$

$$N_{1/4} = \frac{\pi}{4\left(\cot\varphi + \varphi - \frac{\pi}{2}\right)} \qquad (6.2.10b)$$

$N_{1/4}, N_{1/3}, N_q, N_c$ are the coefficients of bearing capacity, which can be determined according to Table 6.2.1.

Table 6.2.1 The value of $N_{1/4}, N_{1/3}, N_q, N_c$ with φ

φ	$N_{1/4}$	$N_{1/3}$	N_q	N_c
0	0	0	1.0	3.14
2	0.06	0.08	1.12	3.32
4	0.12	0.16	1.25	3.51
6	0.20	0.27	1.40	3.71
8	0.28	0.37	1.55	3.93

to be continued

φ	$N_{1/4}$	$N_{1/3}$	N_q	N_c
10	0.36	0.48	1.73	4.17
12	0.46	0.60	1.94	4.42
14	0.60	0.80	2.17	4.70
16	0.72	0.96	2.43	5.00
18	0.86	1.15	2.72	5.31
20	1.00	1.33	3.10	5.66
22	1.20	1.60	3.44	6.04
24	1.40	1.86	3.87	6.45
26	1.60	2.13	4.37	6.90
28	2.00	2.66	4.93	7.40
30	2.40	3.20	5.60	7.95
32	2.80	3.73	6.35	8.55
34	3.20	4.26	7.20	9.22
36	3.60	4.80	8.25	9.97
38	4.20	5.60	9.44	10.80
40	5.00	6.66	10.84	11.73
42	5.80	7.73	12.70	12.80
44	6.40	8.52	14.50	14.00
45	7.40	9.86	15.60	14.60

6.3 Prandtl's bearing capacity theory

Prandtl (1920) showed theoretically that a wedge of material is trapped below a rigid strip footing when it is subjected to concentric loads. A suitable failure mechanism for a strip footing is shown in Fig. 6.3.1. The footing, with width B and infinite length, carries a uniform pressure p on the surface of a mass of homogeneous, isotropic soil. The shear strength parameters for the soil are c and

φ, but the unit weight is assumed to be zero.

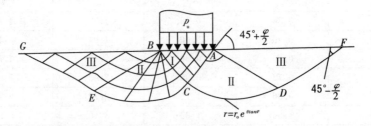

Fig. 6.3.1   General failure under strip footing (Prandtl)

When the pressure becomes equal to the ultimate bearing capacity q_f, the state of plastic equilibrium, in the form of an active Rankine zone is below the footing, the angles ABC and BAC being $(45° + \frac{\varphi}{2})$. The downward movement of the wedge ABC forces the adjoining soil sideways, producing outward lateral forces on both sides of the wedge. Passive Rankine zones ADF and BGE therefore develop on both sides of the wedge ABC, the angles DFA and EGB being $(45° - \frac{\varphi}{2})$. The transition between the downward movement of the wedge ABC and the lateral movement of the wedges ADF and BGE takes place through zones of radial shear (also known as slip fans) CDF and CEG, the surfaces EC and CD being logarithmic spirals (or circular arcs if $\varphi = 0$) to which AC and GE, or BC and FD, are tangential. A state of plastic equilibrium thus exists above the surface GECDF, the remainder of the soil mass being in a state of elastic equilibrium.

The following exact solution can be obtained, using plasticity theory, for the ultimate bearing capacity of a strip footing in the surface of a weightless soil.

$$p_u = c\left[e^{\pi \tan \varphi} \cdot \tan^2\left(45° + \frac{\varphi}{2}\right) - 1 \right] \cdot \cot \varphi = c \cdot N_c \qquad (6.3.1)$$

where

$$N_c = \left[e^{\pi \tan \varphi} \cdot \tan^2\left(45° + \frac{\varphi}{2}\right) - 1 \right] \cdot \cot \varphi \qquad (6.3.2)$$

N_c is the coefficient of bearing capacity, which depends on the friction angle φ and can be obtained from Table 6.3.1.

Table 6.3.1 The coefficient of bearing capacity

$\varphi°$	N_γ	N_q	N_c	$\varphi°$	N_γ	N_q	N_c
0	0	1.00	5.14	24	6.90	9.61	19.3
2	0.01	1.20	5.69	26	9.53	11.9	22.3
4	0.05	1.43	6.17	28	13.1	14.7	25.8
6	0.14	1.72	6.82	30	18.1	18.4	30.2
8	0.27	2.06	7.52	32	25.0	23.2	35.5
10	0.47	2.47	8.35	34	34.5	29.5	42.2
12	0.76	2.97	9.29	36	48.1	37.8	50.6
14	1.16	3.58	10.4	38	67.4	48.9	61.4
16	1.72	4.33	11.6	40	95.5	64.2	75.4
18	2.49	5.25	13.1	42	137	85.4	93.7
20	3.54	6.40	14.8	44	199	115	118
22	4.96	7.82	16.9	45	241	134	133

6.4 Modification of Prandtl's bearing capacity theory

(1) Modification of Reissner

For the general case in which the shear strength parameters are c and φ it is necessary to consider a surcharge pressure q acting on the soil surface would be zero. For this case Reissner (1924) presented the solution due to the surcharge pressure q:

$$p_u = qe^{\pi\tan\varphi} \cdot \tan^2\left(45° + \frac{\varphi}{2}\right) = qN_q \tag{6.4.1}$$

where

$$N_q = e^{\pi\tan\varphi}\tan^2\left(45° + \frac{\varphi}{2}\right) \tag{6.4.2}$$

N_q is the coefficient of bearing capacity, and can be obtained from Table 6.3.1.

Combining Eq. (6.3.1) and Eq. (6.4.1), the ultimate bearing capacity is:
$$p_u = \gamma_0 d N_q + c N_c \qquad (6.4.3)$$

(2) Modification of Taylor

However, an additional term must be added to Eq. (6.4.3) to take the component of bearing capacity due to the self-weight of the soil into account. Taylor (1940) proposed that if the gravity of soil is taken into account and the failure surface is the same as Prandtl's assumption, the shear strength increased accordingly. Taylor assumed that the incremental shear strength can be denoted by an equivalent cohesive force $c' = \gamma t \cdot \tan\varphi$ in which t is the height of the sliding failure soil. So replacing c with $c + c'$ in Eq. (6.4.3), the ultimate bearing capacity considering the self-weight of soil is:

$$\begin{aligned} p_u &= q N_q + (c + c') N_c = q N_q + c' N_c + c N_c \\ &= q N_q + c N_c + \gamma \cdot \frac{B}{2} \tan\left(45° + \frac{\varphi}{2}\right)\left[e^{\pi\tan\varphi} \tan^2\left(45° + \frac{\varphi}{2}\right) - 1\right] \\ &= \frac{1}{2} \gamma \cdot B N_\gamma + q N_q + c N_c \end{aligned}$$

$$(6.4.4)$$

where

$N_\gamma = \tan\left(45° + \frac{\varphi}{2}\right)\left[e^{\pi\tan\varphi} \tan^2\left(45° + \frac{\varphi}{2}\right) - 1\right]$ is obtained from Table 6.3.1.

6.5 Terzaghi's bearing capacity theory

Terzaghi (1943) derived bearing capacity equations based on Prandtl (1920) failure mechanism and the limit equilibrium method for a footing at a depth d below the ground level of a homogeneous soil. The soil is a semi-infinite, foundations are not normally located on the surface of a soil mass, as assumed in the above solutions; but at a depth d below the surface as shown in Fig. 6.5.1. Terzaghi assumed the following:

(1) The soil is a semi-infinite, homogeneous, isotropic, weightless, rigid-plastic material.

(2) The embedment depth is not greater than the width of the footing ($d <$ B).

(3) General shear failure occurs.

(4) The angle α in the wedge (Fig. 6.5.1) is φ. Later, it was found (Vesic, 1973) that $\alpha = 45° + \varphi/2$.

(5) The shear strength of the soil above the footing base is negligible.

(6) The soil above the footing base can be replaced by a surcharge stress (= $\gamma_0 d$).

(7) The base of the footing is rough.

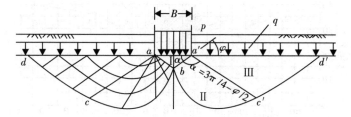

Fig. 6.5.1   Failure surface assumed by Terzaghi

The ultimate bearing capacity of the soil under a shallow strip footing can be expressed by the following general equation

$$p_u = \frac{1}{2}\gamma B N_\gamma + c N_c + q N_q \qquad (6.5.1)$$

where N_γ, N_c and N_q are bearing capacity factors depending only on the value of φ. The first term in Eq. (6.5.1) represents the contribution to bearing capacity resulting from the self-weight of the soil, the second term is the contribution due to the constant component of shear strength and the third term is the contribution due to the surcharge pressure. It should be realized, however, that the superposition of the components of bearing capacity is theoretically incorrect for a plastic material. However, any resulting error is considered to be on the safe side.

For many years Terzaghi's bearing capacity factors were widely used. Terzaghi assumed that the angles baa' and $a'ab$ in Fig. 6.5.1 were equal to φ (i.e. aba' was not considered to be an active Rankine zone). Values of N_γ were obtained by determining the total passive resistance and adhesion force on the planes ab and $a'b$. Terzaghi's values of N_c and N_q were obtained by modifying the prandtl-Reissner solution. However, Terzaghi's values have now been largely supersed.

It is now considered that the values of N_c and N_q expressed in Eq. (6.5.1) should be used in bearing capacity calculations, i.e.

$$N_q = \frac{e^{(\frac{3}{2}\pi - \varphi)\tan\varphi}}{2\cos^2(45° + \frac{\varphi}{2})} \quad (6.5.2a)$$

$$N_c = (N_q - 1)\cot\varphi \quad (6.5.2b)$$

For N_γ, Terzaghi did not give the solution, but plotted N_γ, N_q, N_c with φ as shown in Fig. 6.5.2. N_γ can be determined according to the solid line.

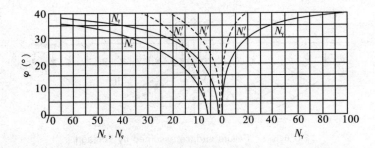

Fig.6.5.2 The coefficient of Terzaghi's bearing capacity

Currently, various equations have been proposed for the N_γ in the literature. Among the popular equations are

Vesic(1973): $\quad N_\gamma = 2(N_q + 1)\tan\varphi_p;$ $\quad (6.5.3a)$

Meyerhof(1976): $\quad N_\gamma = (N_q - 1)\tan 1.4\varphi_p$ $\quad (6.5.3b)$

Davis and Booker (1971):

$$N_\gamma = 0.1054\exp(9.6\varphi_p) \quad \text{for rough footing} \quad (6.5.3c)$$

$$N_\gamma = 0.0663\exp(9.6\varphi_p) \quad \text{for smooth footing} \quad (6.5.3d)$$

The differences between these popular bearing capacity factors are shown in Fig. 6.5.3.

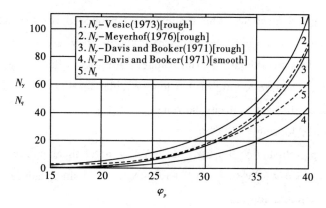

Fig.6.5.3 Comparison of some bearing capacity factors

The bearing capacity factor, N_γ, proposed by Davis and Booker (1971) is based on a refined plasticity method and gives conservative values compared with Vesic (1973). Meyerhof's N_γ values are equal to Davis and Booker's N_γ for φ_p less than about 35°.

Equation (6.5.1) was derived under the condition of the general shear failure, which is suitable for small compressibility soils. For loose and large compressibility soil, the local shear failure may occur, in which the settlement is larger and the ultimate bearing capacity is small. For the case, Eq. (6.5.1) can be modified by reducing the shear strength indexes φ, c.

$$c' = \frac{2}{3}c \tag{6.5.4a}$$

$$\tan\varphi' = \frac{2}{3}\tan\varphi \tag{6.5.4b}$$

Then the ultimate bearing capacity for local shear failure is

$$p_u = c'N'_c + qN'_q + \frac{1}{2}\gamma B N'_\gamma \tag{6.5.5}$$

where N_q', N_γ' and N_c' are obtained according to the dashed line in Fig. 6.5.2.

The problems involved in extending the two-dimensional solution for a strip footing to three dimensions would be considerable. Accordingly, the ultimate bearing capacities of square, rectangular and circular footings are determined by means of semi-empirical shape factors applied to the Solution for a strip footing. The bearing capacity factors N_γ, N_c and N_q should be multiplied by the respective shape factors S_γ, S_c and S_q. The shape factors proposed by Terzaghi and Peck (1996) are still widely used in practice although they are considered to give conservative values of ultimate bearing capacity for high values of φ. The factors are $S_\gamma = 0.8$ for a square footing or 0.6 for a circular footing, $S_c = 1.2$ and $S_q = 1$. Thus the ultimate bearing capacity becomes

For a square footing

$$p_{us} = 0.4\gamma B N_\gamma + \gamma_0 d N_q + 1.2 c N_c \quad \text{general shear failure} \quad (6.5.6)$$

$$p_{us} = 0.4\gamma B N_\gamma' + \gamma_0 d N_q' + 1.2 c' N_c' \quad \text{local shear failure} \quad (6.5.7)$$

and for a circular footing

$$p_{ur} = 0.3\gamma R N_\gamma + \gamma_0 d N_q + 1.2 c N_c \quad \text{general shear failure} \quad (6.5.8)$$

$$p_{ur} = 0.3\gamma R N_\gamma' + \gamma_0 d N_q' + 1.2 c' N_c' \quad \text{local shear failure} \quad (6.5.9)$$

It should be recognized that the results of bearing capacity calculations are very sensitive to the values assumed for the shear strength parameters, especially the higher values of φ. Due consideration must therefore be given to the probable degree of accuracy of the shear strength parameters employed.

Vesic (1973) considered the self-weight of soil and the embedment depth of footing based on Prandtl' theory, and given the bearing capacity for strip footing:

$$p_u = c N_c + q N_q + \frac{1}{2} \gamma B N_\gamma \quad (6.5.10)$$

where N_c, N_q, N_γ are given by, respectively,

$$N_q = e^{\pi \tan\varphi} \tan^2(45° + \frac{\varphi}{2}) \quad (6.5.11a)$$

$$N_c = (N_q - 1) \cdot \cot\varphi \quad (6.5.11b)$$

$$N_\gamma = 2(N_q + 1) \tan\varphi \quad (6.5.11c)$$

According to the factors influencing the bearing capacity, Vesic modified Eq. (6.5.10) and proposed many formulas of bearing capacity.

Equation (6.5.10) is only suitable for strip footing. For other shape footings, the following equation can be used

$$p_u = c N_c S_c + q N_q S_q + \frac{1}{2} \gamma B N_\gamma S_\gamma \quad (6.5.12)$$

where S_c, S_q, S_γ are the coefficients of shape.

For rectangle footing

$$\begin{cases} S_c = 1 + \dfrac{B}{l} \cdot \dfrac{N_q}{N_c} \\ S_q = 1 + \dfrac{B}{l} \cdot \tan\varphi \\ S_\gamma = 1 - 0.4\dfrac{B}{l} \end{cases} \quad (6.5.13a)$$

For circle and square footing

$$\begin{cases} S_c = 1 + \dfrac{N_q}{N_c} \\ S_q = 1 + \tan\varphi \\ S_\gamma = 0.60 \end{cases} \quad (6.5.13b)$$

As subjected to the eccentric load, for strip footing, one should replace the width B by $B' = B - 2e$ (e is eccentricity); for rectangle footing, replace the area A by $A' = B'l'$, in which $B' = B - 2e_b$, $l' = l - 2e_l$, e_b and e_l are the eccentricities of load at short and long sides, respectively.

As the footing subjected the eccentric and inclined load, the bearing capacity is given by

$$p_u = cN_c S_c i_c + qN_q S_q i_q + \dfrac{1}{2}\gamma B N_\gamma S_\gamma i_\gamma \quad (6.5.14)$$

in which i_c, i_q and i_γ are load inclination factors.

$$i_c = \begin{cases} 1 - \dfrac{mH}{B'l'cN_c} & (\varphi = 0) \\ i_q - \dfrac{1 - i_q}{N_c \tan\varphi} & (\varphi > 0) \end{cases} \quad (6.5.15a)$$

$$i_q = \left(1 - \dfrac{H}{Q + B'l'c \cdot \cot\varphi}\right)^m \quad (6.5.15b)$$

$$i_\gamma = \left(1 - \dfrac{H}{Q + B'l'c \cdot \cot\varphi}\right)^{m+1} \quad (6.5.15c)$$

where Q is vertical component of inclined loads, (kN);

H is vertical component of inclined loads, (kN);

B' is effective width of footing, (m);

l' is effective length of footing, (m);

m is constant. For strip footing, $m = 2$; as load inclined to short side of footing, $m_b = \dfrac{2+(B/l)}{1+(B/l)}$; inclined to long side of footing, $m_l = \dfrac{2+(l/B)}{1+(l/B)}$; and inclined arbitrary, $m_n = m_l \cos^2 \theta_n + m_b \sin^2 \theta_n$, θ_n is the inclined angle of load(°).

If the shear strength of the soil above the footing base is considered, the bearing capacity is given by

$$p_u = cN_c S_c i_c d_c + qN_q S_q i_q d_q + \dfrac{1}{2}\gamma B N_\gamma S_\gamma i_\gamma d_\gamma \qquad (6.5.16)$$

where d_c, d_q and d_γ are modified coefficient of embedment depth of footing.

$$d_q = \begin{cases} 1 + 2\tan\varphi(1-\sin\varphi)^2 (d/B) & (d \leqslant B) \\ 1 + 2\tan\varphi(1-\sin\varphi)^2 \tan^{-1}(d/B) & (d > B) \end{cases} \qquad (6.5.17a)$$

$$d_c = \begin{cases} 1 + 0.4 d/B & (\varphi = 0, d \leqslant B) \\ 1 + 0.4\tan^{-1}(d/B) & (\varphi = 0, d > B) \\ d_q - \dfrac{1-d_q}{N_c \tan\varphi} & (\varphi > 0) \end{cases} \qquad (6.5.17b)$$

$$d_\gamma = 1 \qquad (6.5.17c)$$

EXERCISES

6.1 A strip footing 1.50m width is located at a depth of 1.0m below the ground surface, The soil profile at the site is plain fill (with 0.8m thickness, $\gamma = 18\text{kN/m}^3$, water content is 35%) and clay (the thickness is 6m, $\gamma = 18.2\text{kN/m}^3$, water content is 38%, $c = 10\text{kPa}$, $\varphi = 13°$). (1) determine the critical edge pressure p_{cr}, $p_{\frac{1}{4}}$ and $p_{\frac{1}{3}}$; (2) assuming the groundwater level is 1.0m below the ground surface and the strength parameters keep unchange, determine p_{cr}, $p_{\frac{1}{4}}$ and $p_{\frac{1}{3}}$.

6.2 A strip footing 1.50m width is located at a depth of 1.20m below the ground surface in the clay ($\gamma = 18.4\text{kN/m}^3$, $\gamma_{sat} = 18.8\text{kN/m}^3$, $c = 8\text{kPa}$, $\varphi = 15°$). According to the Terzaghi's bearing capacity, (1) determine the bearing capacity of soil with the general shear failure, and the allowable bearing capacity assuming a factor of safety of 2.5; (2) compare the bearing capacity when the footing is located at the depth of 1.6m and 2.0m, respectively. $N_c = 12.9$, $N_q = 4.45$, $N_\gamma = 1.8$.

REFERENCES

1. Liao Hongjian, Liu Houxiang. Soil Mechanics[M]. Beijing: Higher Education Press, 2013.
2. Budhu M. Soil Mechanics and Foundation[M]. John Wiley & Sons, Inc, 2007
3. Craig R F. Craig's Soil Mechanics, (the seventh edition) [M]. CRC Press, 2004.
4. DeBeer E E. Experimental Determination of The Shape Factors and The Bearing Capacity Factors of Sand[J]. Geotechnique, 1970, 20(4): 347 −411.
5. Meyerhof GG, Koumoto T. Inclination Factors for Bearing Capacity of Shallow Footings[J]. J. Geotech. Eng. Div. ASCE, 1987, 113(9): 1013 −1018.
6. Skempton, AW. The Bearing Capacity of Clay [M]. Building Research Congress, London, 1951.
7. Wood DM. Soil Behavior and Critical State Soil Mechanics[M]. Cambridge University Press, Cambridge, 1990.
8. Zadroga, B. Bearing Capacity of Shallow Foundations on Noncohesive Soils [J]. J. Geotech. Eng. , 1994, 120(11): 1991 −2008.
9. Terzaghi K, Peck RB, Mesri G. Soil Mechanics in Engineering Practice (the third edition)[M]. John Wiley & Sons, Inc, 1996.

SOIL MECHANICS
CHAPTER 7
Stability of slopes

7.1 Introduction

Gravitational and seepage forces tend to cause instability in natural, cut and fill slopes and in the slopes of embankments. The most important types of slope failure are illustrated in Fig. 7.1.1. In rotational slips the shape of the failure surface in section may be a circular arc or a non-circular curve. In general, circular slips are associated with homogeneous, isotropic soil conditions and non-circular slips with non-homogeneous conditions. Translational and compound slips occur where the form of the failure surface is influenced by the presence of an adjacent stratum of significantly different strength, most of the failure surface being likely to pass through the stratum of lower shear strength. The form of the surface would also be influenced by the presence of discontinuities such as fissures and pre-existing slips. Translational slips tend to occur where the adjacent stratum is at a relatively shallow depth below the surface of the slope, the failure surface tending to be plane and roughly parallel to the slope. Compound slips usually occur where the adjacent stratum is at greater depth, the failure surface consisting of curved and plane sections. In most cases, slope stability can be considered as a two-dimensional problem, conditions of plane strain being assumed.

Design is based on the requirement to maintain stability rather than on the need to minimize deformation. If deformation were such that the strain in an element of soil exceeded the value corresponding to peak strength, then the strength would fall towards the ultimate value. Thus it is appropriate to use the critical-state strength in analyzing stability. However, if a pre-existing slip surface were to be present within the soil, use of the residual strength would be appropriate. Limiting equilibrium methods are normally used in the analysis of slope stability in which it is considered that failure is on the point of occurring along an assumed or a known failure surface. In the traditional approach the shear strength required to maintain a

Chapter 7 Stability of slopes

condition of limiting equilibrium is compared with the available shear strength of the soil, giving the average (lumped) factor of safety along the failure surface.

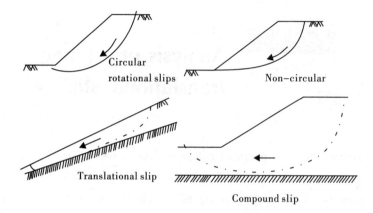

Fig. 7.1.1 Types of slope failure

Alternatively, the limit state method can be used in which partial factors are applied to the shear strength parameters. The ultimate limit state of overall stability is then satisfied if, depending on the method of analysis, either the design disturbing force is less than or equal to the design resisting force along the potential failure surface or the design disturbing moment is less than or equal to the design resisting moment.

The following limit states should be considered as appropriate:
(1) Loss of overall stability due to slip failure.
(2) Bearing resistance failure below embankments.
(3) Internal erosion due to high hydraulic gradients and/or poor compaction.
(4) Failure as a result of surface erosion.
(5) Failure due to hydraulic uplift.
(6) Excessive soil deformation resulting in structural damage to, or loss of service-ability of, adjacent structures, highways or services.

7.2 Analysis of a plane translational slip

It is assumed that the potential failure surface is parallel to the surface of the slope and is at a depth that is small compared with the length of the slope. The slope can then be considered as being of infinite length, with end effects being ignored. The slope is inclined at angle β to the horizontal and the depth of the failure plane is z, as shown in Fig. 7.2.1. The water table is taken to be parallel to the slope at a height of mz ($0 < m < 1$) above the failure plane. Steady seepage is assumed to be taking place in a direction parallel to the slope. The forces on the sides of any vertical slice are equal and opposite, and the stress conditions are the same at every point on the failure plane.

In terms of effective stress, the shear strength of the soil along the failure plane (using the critical-state strength) is

$$\tau_f = (\sigma - u) \tan \varphi'_{cu}$$

Fig. 7.2.1 Plane translational slip

And the factor of safety is

$$F = \frac{\tau_f}{\tau}$$

The expressions for σ, τ and u are

$$\sigma = \{(1-m)\gamma + m\gamma_{sat}\}z\cos^2\beta$$
$$\tau = \{(1-m)\gamma + m\gamma_{sat}\}z\sin\beta\cos\beta$$
$$u = mz\gamma_w \cos^2\beta$$

If the soil between the surface and the failure plane is not fully saturated (i. e. $m=0$) then

$$F = \frac{\tan\varphi'_{cu}}{\tan\beta} \qquad (7.2.1)$$

If the water table coincides with the surface of the slope (i. e. $m=1$) then

$$F = \frac{\gamma'\tan\varphi'_{cu}}{\gamma_{sat}\tan\beta} \qquad (7.2.2)$$

For a total stress analysis the shear strength parameter c_u is used (with $\varphi_u = 0$) and the value of u is zero.

Example 7.1

A long natural slope in overconsolidated fissured clay of saturated unit weight 20kN/m^3 is inclined at $12°$ to the horizontal. The water table is at the surface and seepage is roughly parallel to the slope. A slip has developed on a plane parallel to the surface at a depth of 5m. Determine the factor of safety along the slip plane using (a) the critical-state parameter $\varphi'_{cu} = 28°$ and (b) the residual strength parameter $\varphi'_r = 20°$.

 olution

Equation 7.2.2 applies in both cases.

(1) In terms of critical-state strength

$$F = \frac{10.2\tan 28°}{20\tan 12°} = 1.28$$

(2) In terms of residual strength

$$F = \frac{10.2\tan 20°}{20\tan 12°} = 0.87$$

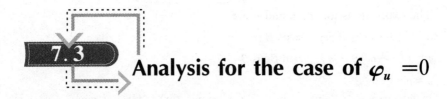

Analysis for the case of $\varphi_u = 0$

7.3.1 Factor of safety for the case of $\varphi_u = 0$ using circular failure surface

A plane strain slope of height H and angle β is shown in Fig. 7.3.1 (a). The trial circular failure surface is defined by its centre C, radius R and central angle θ. Shear stresses along the trial surface are due only to undrained cohesion and are mobilized to c_u/F (to maintain the equilibrium of the sliding block), where F is the factor of safety. The weight of the sliding block W acts at a distance d from the centre of the circle. Taking moments of the forces about the centre of the circular arc, and noting that the normal stresses on the arc pass through the centre, then:

$$W \times d - \int_0^\theta dl \times 1.0 \frac{c_u}{F} \times R = W \times d - \frac{c_u R}{F}\int_0^\theta dl = W \times d - \frac{c_u R^2 \theta}{F} = 0,$$

thus:

$$F = \frac{c_u R^2 \theta}{Wd} = \frac{c_u L_a R}{Wd} \qquad (7.3.1)$$

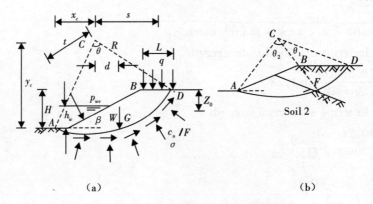

Fig. 7.3.1　Slope failure in $\varphi_u = 0$ soil

where L_α is the length of the circular arc. If a surcharge loading q is applied to the upper-ground surface and external water exists at the front of the slope (Fig. 7.3.1 (a)):

$$F = \frac{c_u R^2 \theta}{Wd + qLs - P_{we}t} \quad (7.3.2)$$

If the soil is composed of two different layers (Fig. 7.3.1 (b)), F is obtained from

$$F = \frac{R^2(c_{u1}\theta'_1 c_{u2}\theta'_2)}{Wd} \quad (7.3.3)$$

where c_{u1} and c_{u2} are the undrained cohesions of soil 1 and soil 2 respectively and θ'_1 and θ'_2 are the corresponding central angles. The stability of an earth slope in undrained conditions can be expressed in terms of a dimensionless parameter N called the stability number

$$N = \gamma H / c_u \quad (7.3.4)$$

It can be shown that for a specified value of β, the magnitude of N at failure has a constant value (N_f) and, as a consequence, the factor of safety Eq. (7.3.1) may be presented by

$$F = N_f / N_d \quad (7.3.5)$$

where N_d is the stability number corresponding to the design values of γ, H and c_u. The centroid of the sliding mass is obtained using a mathematical procedure based on the geometry or the sub-division of the sliding mass into narrow vertical slices.

Example 7.2

Find the factor of safety of a 1 vertical to 1.5 horizontal slope that is 6 m high. The centre of the trial circle is located 2.5 m to the right of and 9.15 m above the toe of the slope. $c_u = 25$ kPa, and $\gamma = 18$ kN/m^3.

Geometrical data are: $\theta = 89.5°$, area of the sliding mass $= 29.87$ m^2 and d $= 3.85$ m.

$R = \sqrt{9.15^2 + 2.5^2} = 9.48 m$, $W = 29.87 \times 1.0 \times 18.0 = 537.7 kN$

$F = c_u R^2 \theta / Wd = [25.0 \times 9.48^2 \times (85.9°/180.0°)\pi]/(537.7 \times 3.85) = 1.63$

7.3.2 Location of the critical circle

Cousins (1978) developed a series of charts for homogeneous soils using extensive computer analyses to investigate the effects of pore pressure and a hard stratum (located horizontally under the slope) on the magnitude of the stability number N. Figure 7.3.2 is reconstructed for a $\varphi_u=0$ soil where the hard layer is deep and has no effect on the mode of failure. The coordinate of the centre of the critical circle through the toe (x_c, y_c) in the xAy plane (with its origin A at the toe) is normalized in terms of height and slope angle. Note that the most critical circle may or may not be a toe circle. Selecting a trial circle using Fig. 7.3.2 or other available charts can reduce the number of iterations in the search for critical circle.

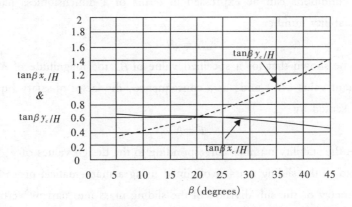

Fig. 7.3.2 The position of the critical toe circle in $\varphi_u=0$ soil (Cousins, 1978)

Example 7.3

A cut 10 m deep is to be made in a stratum of cohesive soil for which $c_u=$ 45kPa. The slope angle $\beta=40°$ and the soil has a unit weight of 17kN/m³. Using Cousins' chart of Fig. 7.3.2, find the factor of safety for the critical toe circle.

From Fig. 7.3.2 $\tan\beta x_c/H \approx 0.50$, $\tan\beta y_c/H \approx 1.21$.

$x_c = 0.5 \times 10.0/\tan 40.0° \approx 6m$, $y_c = 1.21 \times 10.0/\tan 40.0° \approx 14.4m$

The central angle, area of the sliding block and the position of its centroid are found to be:

$\theta = 96.2°$, $S = 128.5 \text{m}^2$, $d = 5.24\text{m}$

$R = \sqrt{6.0^2 + 14.4^2} = 15.6\text{m}$, $W = 128.5 \times 1.0 \times 17.0 = 2184.5 \text{kN}$.

From Equation 7.3.1

$F = (45.0 \times 15.6^2 \times 96.2° \times \pi/180°)/(2184.5 \times 5.24) = 1.60$.

7.3.3 Taylor's stability charts

In undrained conditions, a horizontal hard stratum located $n_d H$ below the upper ground surface affects the critical stability number N_f. The stability number increases as n_d decreases. For $\beta > 53°$, the critical circle is a toe circle and the hard stratum has no effect on the stability number. The slope stability chart developed by Taylor (1948) is shown in Fig. 7.3.3, where the dashed curves represent the undrained conditions.

Example 7.4

Using Taylor's stability chart of Fig. 7.3.3, determine the factor of safety for the slope of Example 7.3.

Solution

For $\beta = 40°$ the stability number from the chart is 5.52.

$N_d = \gamma H/c_u = 17.0 \times 10.0/45.0 = 3.78$, $F = 5.52/3.78 = 1.46$. This is less than the 1.60 obtained in Example 7.3, indicating that the selected circular surface was not the critical one.

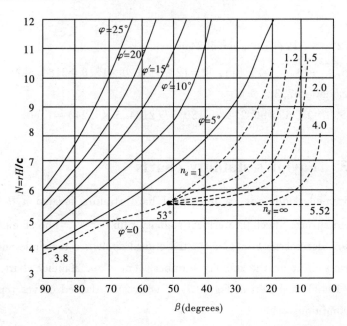

Fig. 7.3.3   Relationship between stability number and slope angle (Taylor, 1948)

The method of slices

◀ 7.4.1   The Fellenius (or Swedish) method of slices

In this method the potential failure surface, in section, is again assumed to be a circular arc with centre O and radius r. The soil mass (ABCD) above a trial failure surface (AC) is divided by vertical planes into a series of slices of width b, as shown in Fig. 7.4.1. The base of each slice is assumed to be a straight line. For any slice the inclination of the base to the horizontal is α and the height, measured on the centre-line, is h. The analysis is based on the use of a lumped factor of safety (F), defined as the ratio of the available shear strength (τ_f) to the shear strength (τ_m) which must be mobilized to maintain a condition of limiting equilibrium, i. e.

$$F = \frac{\tau_f}{\tau_m}$$

The factor of safety is taken to be the same for each slice, implying that there must be mutual support between slices, i. e. forces must act between the slices.

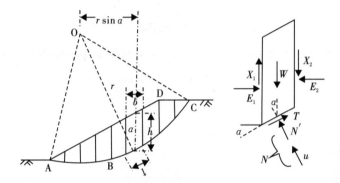

Fig. 7.4.1 The method of slices

The forces (per unit dimension normal to the section) acting on a slice are:
(1) The total weight of the slice, $W = \gamma b h$ (γ_{sat} where appropriate).
(2) The total normal force on the base, N (equal to σl). In general this force has two components, the effective normal force N' (equal to $\sigma' l$), and the boundary water force U (equal to ul), where u is the pore water pressure at the centre of the base and l the length of the base.
(3) The shear force on the base, $T = \tau_m l$.
(4) The total normal forces on the sides, E_1 and E_2.
(5) The shear forces on the sides, X_1 and X_2.
Any external forces must also be included in the analysis.

The problem is statically indeterminate and in order to obtain a solution assumptions must be made regarding the interslice forces E and X; in general the resulting solution for factor of safety is not exact.

Considering moments about O, the sum of the moments of the shear forces T on the failure arc AC must equal the moment of the weight of the soil mass ABCD. For any slice the lever arm of W is $r\sin\alpha$, therefore

$$\sum T_r = \sum W_r \sin\alpha$$

••• 171

Now

$$T = \tau_m l = \frac{\tau_f}{F} l$$

$$\therefore \sum \frac{\tau_f}{F} = \sum W\sin\alpha$$

$$\therefore F = \frac{\sum \tau_f l}{\sum W\sin\alpha}$$

For an effective stress analysis (in terms of tangent parameters c' and φ')

$$F = \frac{\sum(c' + \sigma'\tan\varphi')}{\sum W\sin\alpha}$$

or

$$F = \frac{c'L_\alpha + \tan\varphi' \sum N'}{\sum W\sin\alpha} \qquad (7.4.1a)$$

where L_α is the arc length AC. Equation 7.4.1 (a) is exact but approximations are introduced in determining the forces N'. For a given failure arc the value of F will depend on the way in which the force N' are estimated.

However, the critical-state strength is normally appropriate in the analysis of slope stability, i.e. $\varphi' = \varphi_{cu}'$ and $c' = 0$, therefore the factor of safety is given by

$$F = \frac{\tan\varphi_{cu}' \sum N'}{\sum W\sin\alpha} \qquad (7.4.1b)$$

The Fellenius (or Swedish) solution

In this solution it is assumed that for each slice the resultant of the interslice forces is zero. The solution involves resolving the forces on each slice normal to the base, i.e.

$$N' = W\cos\alpha - ul$$

Hence the factor of safety in terms of effective stress (Eq. 7.4.1 (a)) is given by

$$F = \frac{c'L_\alpha + \tan\varphi' \sum(W\cos\alpha - ul)}{\sum W\sin\alpha} \qquad (7.4.2)$$

The components $W\cos\alpha$ and $W\sin\alpha$ can be determined graphically for each slice. Alternatively, the value of α can be measured or calculated. Again, a series of trial failure surfaces must be chosen in order to obtain the minimum factor of safety. This solution underestimates the factor of safety; the error, compared with more accurate methods of analysis, is usually within the range 5—20%.

For an analysis in terms of total stress the parameter c_u is used in Eq. 7.4.1(a) (with $\varphi_u = 0$) and the value of u is zero. The factor of safety then becomes

$$F = \frac{c_u L_\alpha}{\sum W \sin \alpha} \quad (7.4.3)$$

As N' does not appear in Equation 7.4.3, an exact value of F is obtained. Use of the Fellenius method is not now recommended in practice.

◀ 7.4.2 The Bishop routine method of slices

In this solution it is assumed that the resultant forces on the sides of the slices are horizontal, i.e.

$$X_1 - X_2 = 0$$

For equilibrium the shear force on the base of any slice is

$$T = \frac{1}{F}(c'l + N'\tan\varphi')$$

Resolving forces in the vertical direction:

$$W = N'\cos\alpha + ul\cos\alpha + \frac{c'l}{F}\sin\alpha + \frac{N'}{F}\tan\varphi'\sin\alpha \quad (7.4.4)$$

$$\therefore N' = \frac{[W - (c'l/F)\sin\alpha - ul\cos\alpha]}{[\cos\alpha + (\tan\varphi'\sin\alpha)/F]}$$

It is convenient to substitute

$$l = b\sec\alpha$$

From Equation 7.4.1(a), after some rearrangement,

$$F = \frac{1}{\sum W \sin\alpha} \sum \left[\{c'b + (W - ub)\tan\varphi'\} \frac{\sec\alpha}{1 + (\tan\alpha \tan\varphi'/F)} \right]$$

$$(7.4.5)$$

Bishop (1955) also showed how non-zero values of the resultant forces $(X_1 - X_2)$ could be introduced into the analysis but this refinement has only a marginal effect on the factor of safety.

The pore water pressure can be related to the total 'fill pressure' at any point by means of the dimensionless pore pressure ratio, defined as

$$r_u = \frac{u}{\gamma h} \qquad (7.4.6)$$

(γ_{sat} where appropriate). For any slice,

$$r_u = \frac{u}{W/b}$$

Hence Equation 7.4.5 can be written as

$$F = \frac{1}{\sum W \sin\alpha} \sum \left[\{c'b + W(1 - r_u)\tan\varphi'\} \frac{\sec\alpha}{1 + (\tan\alpha\tan\varphi'/F)} \right] \qquad (7.4.7)$$

As the factor of safety occurs on both sides of Eq. (7.4.7), a process of successive approximation must be used to obtain a solution but convergence is rapid. Due to the repetitive nature of the calculations and the need to select an adequate number of trial failure surfaces, the method of slices is particularly suitable for solution by computer. More complex slope geometry and different soil strata can be introduced.

In most problems the value of the pore pressure ratio r_u is not constant over the whole failure surface but, unless there are isolated regions of high pore pressure, an average value (weighted on an area basis) is normally used in design. Again, the factor of safety determined by this method is an underestimate but the error is unlikely to exceed 7% and in most cases is less than 2%.

Spencer (1967) proposed a method of analysis in which the resultant interslice forces are parallel and in which both force and moment equilibrium are satisfied. Spencer showed that the accuracy of the Bishop routine method, in which only moment equilibrium is satisfied, is due to the insensitivity of the moment equation to the slope of the interslice forces.

Dimensionless stability coefficients for homogeneous slopes, based on Eq. (7.4.7), have been published by Bishop and Morgenstern (1960). It can be shown that for a given slope angle and given soil properties the factor of safety varies linearly with r_u and can thus be expressed as

$$F = m - nr_u \qquad (7.4.8)$$

where m and n are the stability coefficients. The coefficients m and n are functions of β, φ', depth factor D and the dimensionless factor $c'/\gamma H$ (which is zero if the critical-state strength is used).

Example 7.5

Using the Fellenius method of slices, determine the factor of safety, in terms of effective stress, of the slope shown in Fig. 7.4.2 for the given failure surface (a) using peak strength parameters $c' = 10$ kN/m² and $\varphi' = 29°$ and (b) using critical-state parameter $\varphi'_{cu} = 31°$. The unit weight of the soil both above and below the water table is 20kN/m³.

S *olution*

(a) The factor of safety is given by Eq. (7.4.2). The soil mass is divided into slices 1.5 m wide. The weight (W) of each slice is given by

$$W = \gamma b h = 20 \times 1.5 \times h = 30\ h\text{kN/m}$$

The height h for each slice is set off below the centre of the base, and the normal and tangential components $h\cos\alpha$ and $h\sin\alpha$, respectively, are determined graphically, as shown in Fig. 7.4.2. Then

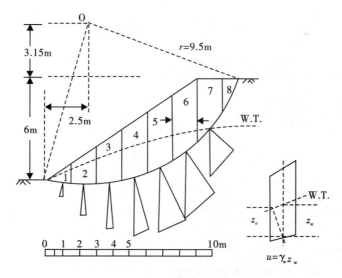

Fig. 7.4.2 Example 7.5

$$W\cos\alpha = 30h\cos\alpha$$

$W\sin\alpha = 30h\sin\alpha$

The pore water pressure at the centre of the base of each slice is taken to be $\gamma_w z_w$, where z_w is the vertical distance of the centre point below the water table (as shown in the figure). This procedure slightly overestimates the pore water pressure which strictly should be $\gamma_w z_e$, where z_e is the vertical distance below the point of intersection of the water table and the equipotential through the centre of the slice base. The error involved is on the safe side.

The arc length (L_α) is calculated as 14.35 m. The results are given in Table 7.4.1.

$$\sum W\cos\alpha = 30 \times 17.50 = 525 \text{kN/m}$$

$$\sum W\sin\alpha = 30 \times 8.45 = 254 \text{kN/m}$$

$$\sum (W\cos\alpha - ul) = 525 - 132 = 393 \text{kN/m}$$

$$F = \frac{c'L_\alpha + \tan\varphi' \sum (W\cos\alpha - ul)}{\sum W\sin\alpha}$$

$$= \frac{(10 \times 14.35) + (0.554 \times 393)}{254}$$

$$= \frac{143.5 + 218}{254} = 1.42$$

Table 7.4.1 Calculation procedure for example 7.5

Slice No.	$h\cos\alpha$(m)	$h\sin\alpha$(m)	u(kN/m^2)	l(m)	ul(kN/m)
1	0.75	−0.15	5.9	1.55	9.1
2	1.80	−0.10	11.8	1.50	17.7
3	2.70	0.40	16.2	1.55	25.1
4	3.25	1.00	18.1	1.60	19.0
5	3.45	1.75	17.1	1.70	29.1
6	3.10	2.35	11.3	1.95	22.0
7	1.90	2.25	0	2.35	0
8	0.55	0.95	0	2.15	0
	17.50	8.45		14.35	132.0

(b) In terms of critical-state strength

$$F = \frac{\tan 31° \times 393}{254} = 0.93$$

Deformation is likely to result in strains along a potential failure surface exceeding the value corresponding to peak strength. Therefore the strength mobilized for stability is likely to fall below the peak value and to approach the critical-state value. Therefore the slope is unsafe. It should be noted that in case (a), the proportion of shear strength represented by c', generally a parameter of uncertain value, is 40%.

7.5 General methods of analysis

Morgenstern and Price (1965) developed a general analysis in which all boundary and equilibrium conditions are satisfied and in which the failure surface may be any shape, circular, non-circular or compound. The ground surface is represented by a function $y = z(x)$ and the trial failure surface by $y = y(x)$ as shown in Fig. 7.5.1. The forces acting on an infinitesimal slice of width dx are also shown in the Figure. The forces are denoted as follows:

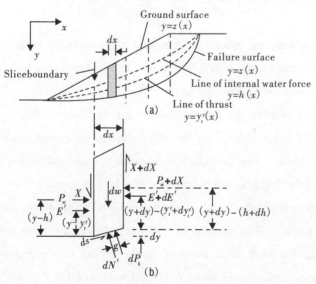

Fig. 7.5.1 The Morgenstern-Price method

E' = effective normal force on a side of the slice,

X = shear force on a side,

P_w = boundary water force on a side,

dN' = effective normal force on the base of the slice,

dS = shear force on the base,

dP_b = boundary water force on the base,

dW = total weight of the slice.

The line of thrust of the effective normal forces (E') is represented by a function $y = y'_t(x)$ and that of the internal water forces (P_w) by $y = h(x)$. Two governing differential equations are obtained by equating moments about the midpoint of the base, and forces perpendicular and parallel to the base, to zero. The equations are simplified by working in terms of the total normal force (E), where

$$E = E' + P_w$$

The position of force E on a side of the slice is obtained from the expression

$$Ey_t = E'y'_t + P_w h$$

The problem is rendered statically determinate by assuming a relationship between the forces E and X of the form

$$X = \lambda f(x) E \qquad (7.5.1)$$

where $f(x)$ is a function chosen to represent the pattern of variation of the ratio X/E across the failure mass and λ is a scale factor. The value of λ is obtained as part of the solution along with the factor of safety F.

To obtain a solution the soil mass above a trial failure surface is divided into a series of slices of finite width such that the failure surface within each slice can be assumed to be linear. The boundary conditions at each end of the failure surface are in terms of the force E and a moment M which is given by the integral of an expression containing both E and X: normally both E and M are zero at each end of the failure surface. The method of solution involves choosing trial values of λ and F, setting the force E to zero at the beginning of the failure surface and integrating across each slice in turn, obtaining values of E, X and y_t: the resulting values of E and M at the end of the failure surface will in general not be zero. A systematic iteration technique, based on the Newton Raphson method and described by Morgenstern and Price (1965), is used to modify the values of λ and F until the resulting values of both E and M at the end of the failure surface are zero. The factor of safety is not significantly affected by the choice of the function $f(x)$ and as a consequence $f(x) = 1$ is a common assumption.

For any assumed failure surface it is necessary to examine the solution to ensure that it is valid in respect of the implied state of stress within the soil mass above that surface. Accordingly, a check is made to ensure that neither shear failure nor a state of tension is implied within the mass. The first condition is satisfied if the available shearing resistance on each vertical interface is greater than the corresponding value of the force X: the ratio of these two forces represents the local factor of safety against shear failure along the interface. The requirement that no tension should be developed is satisfied if the line of thrust of the E forces, as given by the computed values of y_t, lies wholly above the failure surface.

Computer software for the Morgenstern-Price analysis is readily available. The method can be fully exploited if an interactive approach, using computer graphics, is adopted.

Bell (1968) proposed a method of analysis in which all the conditions of equilibrium are satisfied and the assumed failure surface may be of any shape. The soil mass is divided into a number of vertical slices and statical determinacy is obtained by means of an assumed distribution of normal stress along the failure surface.

Sarma (1973) developed a method, based on the method of slices, in which the critical earthquake acceleration required to produce a condition of limiting equilibrium is determined. An assumed distribution of vertical interslice forces is used in the analysis. Again, all the conditions of equilibrium are satisfied and the assumed failure surface may be of any shape. The static factor of safety is the factor by which the shear strength of the soil must be reduced such that the critical acceleration is zero.

The use of a computer is also essential for the Bell and Sarma methods and all solutions must be checked to ensure that they are physically acceptable.

EXERCISES

7.1 For the given failure surface, determine the factor of safety in terms of total stress for the slope detailed in Fig. 7 −1. The unit weight for both soils is $19 kN/m^3$. The characteristic undrained strength (c_{uk}) is 30 kN/m^2 for soil 1 and $45 kN/m^2$ for soil 2. What is the factor of safety if allowance is made for the development of a tension crack? Check the stability of the slope using the limit state method.

7.2 Find the factor of safety of a 1 vertical to 1.5 horizontal slope that is 8 m high. The centre of the trial circle is located 5 m to the right of and 12 m above the toe of the slope. $c_u = 20$ kPa, and $\gamma = 19$ kN/m^3.

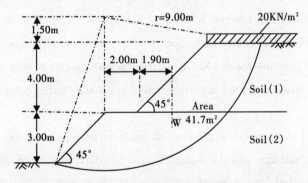

Fig. 7.1 Exercise 7.1

7.3 For the given failure surface, determine the factor of safety in terms of effective stress for the slope detailed in Fig. 7 −2, using the Fellenius method of slices. The unit weight of the soil is $21 kN/m^3$ and the characteristic shear strength parameters are $c' = 8 kN/m^2$ and $\phi' = 32$ °. According to the limit state method, will a slip failure occur?

(Reproduced from Skempton and Brown (1961). A landslide in boulder clay at Selset, Yorkshire, Geotechnique. 11, p. 280, by permission of the Council of the institution of Civil Engineers)

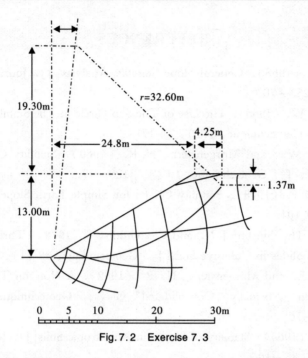

Fig. 7.2 Exercise 7.3

7.4 Using the Bishop routine method of slices determine the factor of safety in terms of effective stress for the slope detailed in Fig. 7.3 for the specified failure surface. The value of r_u is 0.20 and the unit weight of the soil is 20kN/m^3. Characteristic values of the shear strength parameters are $c'=0$ and $\varphi'=33°$.

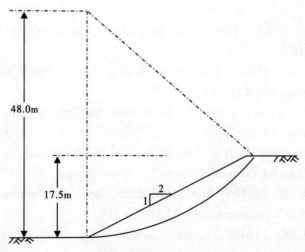

Fig. 7.3 Exercise 7.4

REFERENCES

1. Bell, J. M. (1968). General Slope Stability Analysis[J]. Journal ASCE, 94 (SM6), 1253 −70.
2. Bishop, A. W. (1955). The Use of The Slip Circle in The Stability Analysis of Slopes[J]. Geotechnique, 5(1), 7 −17.
3. Bishop, A. W., and Morgenstern, N. R. (1960). Stability Coefficients for Earth Slopes[J]. Geotechnique, 10 (4), 129 −47.
4. Cousins, B. F. (1978). Stability Charts for Simple Earth Slopes[J]. Journal GE, ASCE 104.
5. Gens, A., Hutchinson, J. N. and Cavounidis, S. (1988). Three-dimensional Analysis of Slides in Cohesive Soils[J]. Geotechnique, 38 (1), 1 −23.
6. Gibson, R. E. and Morgenstern, N. R. (1962). A Note on The Stability of Cuttings in Normally Consolidated Clays, Geotechnique [J]. 12 (3), 212 −16.
7. Lo, K. Y. (1965). Stability of Slopes in Anisotropic Soils[J]. Journal ASCE, 91 (SM4), 85 −106.
8. Morgenstern, N. R. and Price, V. E. (1965). The Analysis of The Stability of General Slip Surfaces[J]. Geotechnique, 15, 79 −93.
9. Morgenstern, N. R. and Price, V. E. (1967). A Numerical Method for Solving The Equations of Stability of General Slip Surfaces[J]. Computer Journal, 9, 388 − 393.
10. Penman, A. D. M. (1986). On the Embankment Dam[J]. Geotechnique, 36, 301 −48.
11. Sarma, S. K. (1973). Stability Analysis of Embankments and Slopes[J]. Geotechnique, 23, 423 −33.
12. Skempton, A. W. (1964). Long-term Stability of Clay Slopes [J]. Geotechnique, 14, 75 −102.
13. Spencer, E. (1967). A Method of Analysis of The Stability of Embankments Assuming Parallel Inter-slice Forces[J]. Geotechnique, 17, 11 −26.
14. Taylor, D. W. (1937). Stability of Earth Slopes[J]. Journal of the Boston Society of Civil Engineers, 24 (3), 337 −386.
15. Taylor, D. W. (1948). Fundamentals of soil mechanics [M]. New York: Wiley.

SOIL MECHANICS CHAPTER 8

Lateral earth pressure and retaining walls

8.1 Introduction

This chapter deals with the magnitude and distribution of lateral pressure between a soil mass and an adjoining retaining wall. A retaining wall is a structure that retains any material (usually earth) and prevents it from collapsing or eroding away. It is designed so that to resist the pressure of the material that it is holding back.

Various types of earth retaining structures are used in civil engineering, the main ones being: 1) Mass construction gravity walls; 2) Reinforced concrete walls; 3) Crib walls; 4) Gabion walls; 5) Sheet pile walls; 6) Diaphragm walls; 7) Reinforced soil walls; 8) Anchored earth walls.

Earth retaining structures are commonly used to support soils and structures to maintain a difference in elevation of the ground surface and are normally grouped into gravity walls or embedded walls.

Conditions of plane strain are assumed, i. e. strains in the longitudinal direction of the structure are assumed to be zero. The rigorous treatment of this type of problem, with both stresses and displacements being considered, would involve a knowledge of appropriate equations defining the stress strain relationship for the soil and the solution of the equations of equilibrium and compatibility for the given boundary conditions. It is possible to determine displacement by means of the finite element method using suitable computer software, provided realistic values of the relevant deformation parameters are available. However, it is the failure condition of the retained soil mass which is of primary interest and in this context, provided a consideration of displacements is not required, it is possible to use the concept of plastic collapse. Earth pressure problems can thus be considered as problems in plasticity.

It is assumed that the stress-strain behaviour of the soil can be represented by the rigid-perfectly plastic idealization, shown in Fig. 8.1.1, in which both yielding and shear failure occur at the same state of stress: unrestricted plastic flow takes place at this stress level. A soil mass is said to be in a state of plastic equilibrium if the shear stress at every point within the mass reaches the value represented by point Y'.

Plastic collapse occurs after the state of plastic equilibrium has been reached in part of a soil mass, resulting in the formation of an unstable mechanism: that part of the soil mass slips relative to the rest of the mass. The applied load system, including body forces, for this condition is referred to as the collapse load.

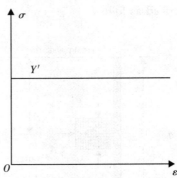

Fig. 8.1.1 Idealized stress–strain relationship

Determination of the collapse load using plasticity theory is complex and would require that the equilibrium equations, the yield criterion and the flow rule were satisfied within the plastic zone. The compatibility condition would not be involved unless specific deformation conditions were imposed. However, plasticity theory also provides the means of avoiding complex analyses. The limit theorems of plasticity can be used to calculate lower and upper bounds to the true collapse load. In certain cases, the theorems produce the same result which would then be the exact value of the collapse load.

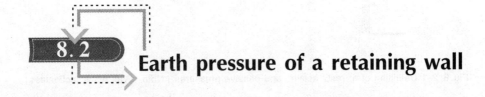

8.2 Earth pressure of a retaining wall

Consider a mass of soil shown in Fig. 8.2.1 (a). The mass is bounded by a frictionless wall of height AB. A soil element located at a depth z is subjected to a vertical effective pressure σ'_0 and a horizontal effective pressure σ'_h. There are no shear stresses on the vertical and horizontal planes of the soil element. Let us define the ratio of σ'_h to σ'_0 as a nondimensional quanity K

$$K = \frac{\sigma'_h}{\sigma'_0} \qquad (8.2.1)$$

Now, three possible cases may arise concerning the retaining wall; and they are described as follows.

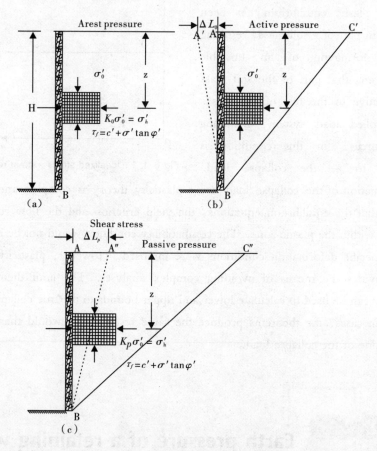

Fig. 8.2.1 Definition of at rest, active, and passive pressures (Note: Wall AB is frictionless)

◀ 8.2.1 At rest earth pressure

If the wall AB is static, that is, if it does not move either to the right or to the

Chapter 8 Lateral earth pressure and retaining walls

left of its initial position, the soil mass will be in a state of static equilibrium. In that case, σ'_h is referred to as the at rest earth pressure, or

$$K = K_0 = \frac{\sigma'_h}{\sigma'_0} \qquad (8.2.2)$$

where K_0 = at rest earth pressure coefficient.

8.2.2 Active earth pressure

If the frictionless wall rotates sufficiently about its bottom to a position of $A'B$ (Fig. 8.2.1(b)), then a triangular soil mass ABC' adjacent to the wall will reach a state of plastic equilibrium and will fail sliding down the plane BC'. At this time, the horizontal effective stress, $\sigma'_h = \sigma'_a$, will be referred to as active earth pressure. Now,

$$K = K_a = \frac{\sigma'_h}{\sigma'_0} = \frac{\sigma'_a}{\sigma'_0} \qquad (8.2.3)$$

where K_a = active earth pressure coefficient.

8.2.3 Passive earth pressure

If the frictionless wall rotates sufficiently about its bottom to a position $A''B$ (Fig. 8.2.1 (c)), then a triangular soil mass ABC'' will reach a state of plastic equilibrium and will fail sliding upward along the plane BC''. The horizontal effective stress at this time will be $\sigma'_h = \sigma'_p$, the so-called passive pressure. In this case,

$$K = K_p = \frac{\sigma'_h}{\sigma'_0} = \frac{\sigma'_p}{\sigma'_0} \qquad (8.2.4)$$

where K_p = passive earth pressure coefficient.

8.3 Earth pressure at rest

The fundamental concept of earth pressure at rest was discussed in the preceding section. In order to define the earth pressure coefficient K_0 at rest, we refer to Fig. 8.3.1, which shows a wall AB retaining a dry soil with a unit weight of γ. The wall is static. At a depth z,

Vertical effective stress $\sigma'_0 = \gamma z$

Horizontal effective stress $\sigma'_h = K_0 \gamma z$

So $K_0 = \dfrac{\sigma'_h}{\sigma'_0} = $ at rest earth pressure coefficient.

For coarse-grained soils, the coefficient of earth pressure at rest can be estimated by using the empirical relationship (Jaky, 1944)

$$K_0 = 1 - \sin \varphi' \qquad (8.3.1)$$

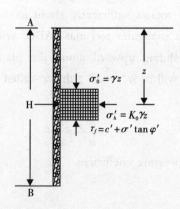

ig. 8.3.1 Earth pressure at rest

where φ' = drained friction angle.

While designing a wall that may be subjected to lateral earth pressure at rest, one must take care in evaluating the value of K_0. Sherif et. al (1984), on the basis of their laboratory tests, showed that Jaky's equation for K_0 (Eq. 8.3.1) gives good results when the back fill is loose sand. However, for a dense sand backfill, Eq. 8.3.1 may grossly underestimate the lateral earth pressure at rest. This underestimation results because of the process of compaction of backfill. For this reason, they recommended the design relationship.

$$K_0 = (1 - \sin \varphi) + \left[\frac{\gamma_d}{\gamma_{d(\min)}} - 1\right] \times 5.5 \qquad (8.3.2)$$

where γ_d = actual compacted dry unit weight of the sand behind the wall

$\gamma_{d(\min)}$ = dry unit weight of the sand in the loosest state

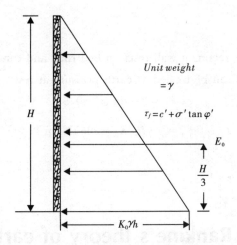

Fig. 8.3.2 Distribution of lateral earth pressure at rest on a wall

For fine-grained, normally consolidated soils, Massarsch (1979) suggested the following equation for K_0

$$K_0 = 0.44 + 0.42 \times \left[\frac{PI(\%)}{100}\right] \qquad (8.3.3)$$

For overconsolidated clays, the coefficient of earth pressure at rest can be approximated as

$$K_{0(\text{overconsolidated})} = K_{0(\text{normally consolidated})} \sqrt{OCR} \qquad (8.3.4)$$

where OCR = overconsolidation ratio, which was defined as

$$OCR = \frac{\text{Preconsolidation pressure}, \sigma'_c}{\text{Present effective overburden pressure}, \sigma'_o} \qquad (8.3.5)$$

Fig. 8.3.2 shows the distribution of lateral earth pressure at rest on a wall of height H retaining a dry soil having a unit weight of γ. The total force per unit length of the wall, E_0, is equal to the area of the pressure diagram, so

$$E_0 = \frac{1}{2}\gamma H^2 K_0 \qquad (8.3.6)$$

The force E_0 acts at a distance of $H/3$ above the bottom of the wall surface.

Example 8.1

A retaining wall rests on bed rock. Height of the wall is $H = 8.0$ m, backfill of the wall is medium dense sand, unit weight and internal friction angle of which are $\gamma = 18.5$ kN/m^3 and $\varphi' = 30°$. Please calculate the earth pressure of the wall.

In this case, the retaining wall rests on bed rock and can not move or rotate, so the earth pressure can be treated as earth pressure at rest

$$K_0 = 1 - \sin\varphi' = 0.5$$

$$E_0 = \frac{1}{2}\gamma H^2 K_0 = 296 \text{kN/m}$$

8.4 Rankine's theory of earth pressure

Rankine's theory (1857) considers the state of stress in a soil mass when the condition of plastic equilibrium has been reached, i.e. when shear failure is on the point of occurring throughout the mass. The theory satisfies the conditions of a lower bound plasticity solution. The Mohr circle representing the state of stress at failure in a two-dimensional element is shown in Fig. 8.4.1, the relevant shear strength parameters being denoted by c and φ. Shear failure occurs along a plane at an angle of $(45° + \frac{\varphi}{2})$ to the major principal plane. If the soil mass as a whole is stressed such that the principal stresses at every point are in the same directions then, theoretically, there will be a network of failure planes (known as a slip line field) equally inclined to the principal planes, as shown in Fig. 8.4.1. It should be appreciated that the state of plastic equilibrium can be developed only if sufficient deformation of the soil mass can take place.

Consider now a semi-infinite mass of soil with a horizontal surface and having a vertical boundary formed by a smooth wall surface extending to semi-infinite depth, as represented in Fig. 8.4.2 (a). The soil is assumed to be homogeneous and isotropic. A soil element at any depth z is subjected to a vertical stress σ_z and a horizontal stress σ_x and, since there can be no lateral transfer of weight if the surface is horizontal, no shear stresses exist on horizontal and vertical planes. The vertical and horizontal stresses, therefore, are principal stresses.

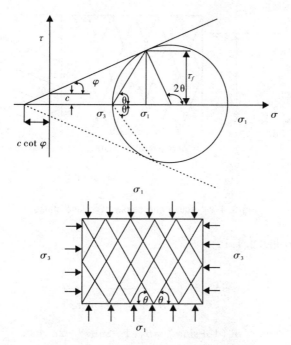

Fig. 8.4.1 State of plastic equilibrium

If there is a movement of the wall away from the soil, the value of σ_x, decreases as the soil dilates or expands outwards, the decrease in σ_x being an unknown function of the lateral strain in the soil. If the expansion is large enough, the value of σ_x decreases to a minimum value such that a state of plastic equilibrium develops. Since this state is developed by a decrease in the horizontal stress σ_x, this must be the minor principal stress (σ_3). The vertical stress σ_z is then the major principal stress (σ_1).

The stress $\sigma_1(=\sigma_z)$ is the overburden pressure at depth z and is a fixed value for any depth. The value of $\sigma_3(=\sigma_x)$ is determined when a Mohr circle through the point representing σ_1 touches the failure envelope for the soil. The relationship between σ_1 and σ_3 when the soil reaches a state of plastic equilibrium can be derived from this Mohr circle. Rankine's original derivation assumed a value of zero for the shear strength parameter c but a general derivation with c greater than zero is given below to cover the cases in which untrained parameter c_u, or tangent parameter c' is used.

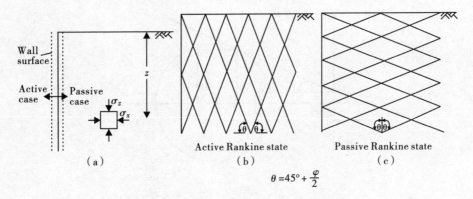

Fig. 8.4.2 Active and passive Rankine states

Referring to Fig. 8.4.1,

$$\sin \varphi = \frac{\frac{1}{2}(\sigma_1 - \sigma_3)}{\frac{1}{2}(\sigma_1 + \sigma_3 + 2c \cot\varphi)}$$

$$\therefore \sigma_3(1 + \sin\varphi) = \sigma_1(1 - \sin\varphi) - 2c \cos\varphi$$

$$\therefore \sigma_3 = \sigma_1 \left(\frac{1 - \sin\varphi}{1 + \sin\varphi}\right) - 2c \left(\frac{\sqrt{1 - \sin^2\varphi}}{1 + \sin\varphi}\right)$$

$$\therefore \sigma_3 = \sigma_1 \left(\frac{1 - \sin\varphi}{1 + \sin\varphi}\right) - 2c \sqrt{\left(\frac{1 - \sin\varphi}{1 + \sin\varphi}\right)} \qquad (8.4.1)$$

Alternatively, $\tan^2(45° - \frac{\varphi}{2})$ can be substituted for $\frac{1 - \sin\varphi}{1 + \sin\varphi}$.

Chapter 8 Lateral earth pressure and retaining walls

As stated, σ_1 is the overburden pressure at depth z, i.e.

$$\sigma_1 = \gamma z$$

The horizontal stress for the above condition is defined as the active pressure (p_a) being due directly to the self-weight of the soil.

If $K_a = \dfrac{1-\sin\varphi}{1+\sin\varphi}$ is defined as the active pressure coefficient, then Eq. (8.4.1) can be written as

$$p_a = K_a \gamma z - 2c\sqrt{K_a} \tag{8.4.2}$$

When the horizontal stress becomes equal to the active pressure the soil is said to be in the active Rankine state, there being two sets of failure planes each inclined at $(45° + \dfrac{\varphi}{2})$ to the horizontal (the direction of the major principal plane) as shown in Fig. 8.4.2(b).

In the above derivation, a movement of the wall away from the soil was considered. On the other hand, if the wall is moved against the soil mass, there will be lateral compression of the soil and the value of σ_x will increase until a state of plastic equilibrium is reached. For this condition, σ_x becomes a maximum value and is the major principal stress σ_1. The stress σ_z, equal to the overburden pressure, is then the minor principal stress, i.e.

$$\sigma_3 = \gamma z$$

The maximum value σ_1 is reached when the Mohr circle through the point representing the fixed value σ_3 touches the failure envelope for the soil. In this case, the horizontal stress is defined as the passive pressure (p_p) representing the maximum inherent resistance of the soil to lateral compression. Rearranging Eq. (8.4.1)

$$\sigma_1 = \sigma_3 \left(\dfrac{1+\sin\varphi}{1-\sin\varphi}\right) + 2c\sqrt{\left(\dfrac{1+\sin\varphi}{1-\sin\varphi}\right)} \tag{8.4.3}$$

If $K_p = \dfrac{1+\sin\varphi}{1-\sin\varphi}$ is defined as the passive pressure coefficient, then Eq. 8.4.3 can be written as

$$p_p = K_p \gamma z + 2c\sqrt{K_p} \tag{8.4.4}$$

When the horizontal stress becomes equal to the passive pressure the soil is said to be in the passive Rankine state, there being two sets of failure planes each inclined at $(45° + \dfrac{\varphi}{2})$ to the vertical (the direction of the major principal plane) as shown in Fig. 8.4.2(c).

Inspection of Eqs. (8.4.2) and (8.4.4) shows that the active and passive pressures increase linearly with depth as represented in Fig. 8.4.3. When $c = 0$, triangular distributions are obtained in each case.

When c is greater than zero, the value of p_a is zero at a particular depth z_0. From Eq. (8.4.2), with $p_a = 0$,

$$z_0 = \dfrac{2c}{\gamma \sqrt{K_a}} \tag{8.4.5}$$

This means that in the active case the soil is in a state of tension between the surface and depth z_0. In practice, however, this tension cannot be relied upon to act on the wall, since cracks are likely to develop within the tension zone and the part of the pressure distribution diagram above depth z_0 should be neglected.

The force per unit length of wall due to the active pressure distribution is referred to as the total active thrust (E_a). For a vertical wall surface of height H

$$E_a = \int_{z_0}^{H} p_a \, dz$$

$$= \dfrac{1}{2} K_a \gamma (H^2 - z_0^2) - 2c \sqrt{K_a} (H - z_0) \tag{8.4.6 a}$$

$$= \dfrac{1}{2} K_a \gamma (H - z_0)^2 \tag{8.4.6 b}$$

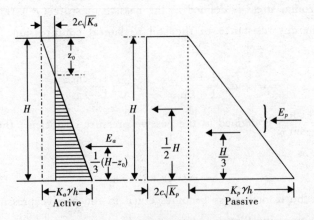

Fig. 8.4.3 Active and passive pressure distributions

The force E_a acts at a distance of $(H-z_0)/3$ above the bottom of the wall surface.

The force due to the passive pressure distribution is referred to as the total passive resistance (E_p). For a vertical wall surface of height H

$$E_p = \int_0^H p_p dz = \frac{1}{2} K_p \lambda H^2 + 2c \sqrt{K_p} H \qquad (8.4.7)$$

The two components of E_p act at distances of $H/3$ and $H/2$, respectively, above the bottom of the wall surface.

If a uniformly distributed surcharge pressure of q per unit area acts over the entire surface of the soil mass, the vertical stress σ_z at any depth is increased to $\gamma_z + q$, resulting in an additional pressure of $K_a q$ in the active case or $K_p q$ in the passive case, both distributions being constant with depth as shown in Fig. 8.4.4. The corresponding forces on a vertical wall surface of height H are $K_a qH$ and $K_p qH$, respectively, each acting at mid-height. The surcharge concept can be used to obtain the pressure distribution in stratified soil deposits. In the case of two layers of soil having different shear strengths, the weight of the upper layer can be considered as a surcharge acting on the lower layer. There will be a discontinuity in the pressure diagram at the boundary between the two layers due to the different values of shear strength parameters.

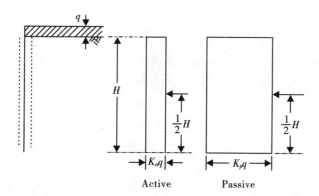

Fig. 8.4.4 Additional pressure due to surcharge

If the soil below the water table is in the fully drained condition, the active and passive pressures must be expressed in terms of the effective weight of the soil and the effective stress parameters c' and φ'. For example, if the water table is at the surface and if no seepage is taking place, the active pressure at depth z is given by

$$p_a = K_a \gamma' z - 2c'\sqrt{K_a}$$

where, $K_a = \dfrac{1-\sin\varphi'}{1+\sin\varphi'}$

Corresponding equations apply in the passive case. The hydrostatic pressure $\gamma_w z$ due to the water in the soil pores must be considered in addition to the active or passive pressure.

For the undrained condition in a fully saturated clay, the active and passive pressures are calculated using the parameter c_u (φ_u being zero) and the total unit weight γ_{sat} (i.e. the water in the soil pores is not considered separately). The effect of the tension zone must be considered for this condition. In theory, a (dry) crack could open to a depth (z_0) of $2c_u/\gamma_{sat}$ (i.e. Eq. (8.4.5) with $K_a = 1$ for $\varphi_u = 0$). Cracking is most likely to occur at the clay/wall interface where the resistance to fracture is lower than that within the clay. If a crack at the interface were to fill with water (due to heavy rainfall or another source of inflow), then hydrostatic pressure would act on the wall. Thus the clay would be supported by the water filling the crack to the depth (z_{0w}) at which the active pressure equals the hydrostatic pressure. Thus, assuming no surface surcharge:

$$\gamma_{sat} z_{0w} - 2c_u = \gamma_w z_{0w}$$

$$\therefore z_{0w} = \frac{2c_u}{(\gamma_{sat} - \gamma_w)}$$

In the Rankine theory it is assumed that the wall surface is smooth whereas in practice considerable friction may be developed between the wall and the adjacent soil, depending on the wall material. In principle, the theory results either in an overestimation of active pressure and an underestimation of passive pressure (i.e. lower bounds to the respective 'collapse loads') or in exact values of active and passive pressures.

Example 8.2

(1) Calculate the total active thrust on a vertical wall 5 m high retaining a sand of unit weight 17 kN/m^3; for which $\varphi' = 35°$; the surface of the sand is horizontal and the water table is below the bottom of the wall. (b) Determine the thrust on the wall if the water table rises to a level 2 m below the surface of the sand. The

saturated unit weight of the sand is 20kN/m³.

 **olution**

$$(a) K_a = \frac{1-\sin 35°}{1+\sin 35°} = 0.27$$

$$E_a = \frac{1}{2}K_a\gamma H^2 = 57.5 \text{kN/m}$$

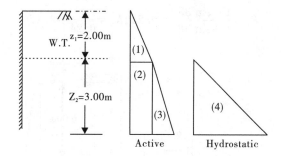

Fig. 8.4.5 Example 8.2

(2) The pressure distribution on the wall is now as shown in Fig. 8.4.5, including hydrostatic pressure on the lower 3 m of the wall.

Above the water table, the active earth pressure is,

$$p_{a1} = K_a\gamma z_1 = 0.27 \times 17 \times 2 = 9.18 \text{ kPa}$$

Below the water table, the active earth pressure should be calculated in terms of effective weight of the soil, so the active earth pressure at the bottom of the wall is,

$$p_{a2} = p_{a1} + K_a\gamma' z_2 = 9.18 + 0.27 \times (20-9.8) \times 3 = 17.44 \text{ kPa}$$

The water pressure is,

$$p_w = \gamma_w z_2 = 9.8 \times 3 = 29.4 \text{ kPa}$$

Therefore the total thrust is (Fig. 8.4.5),

$$E_a = \frac{1}{2} \times 9.18 \times 2 + \frac{1}{2} \times (9.18+17.44) \times 3 + \frac{1}{2} \times 29.4 \times 3 = 93.21 \text{ kN/m}$$

Example 8.3

The soil conditions adjacent to a sheet pile wall are given in Fig. 8.4.6, a surcharge pressure of 50kN/m² being carried on the surface behind the wall. For soil 1, a sand above the water table, $c'=0$, $\varphi'=38°$ and $\gamma=18\text{kN/m}^3$. For soil 2, a saturated clay, $c'=10\text{kN/m}^2$, $\varphi'=28°$ and $\gamma_{sat}=20\text{ kN/m}^3$. Plot the distributions of active pressure behind the wall and passive pressure in front of the wall.

For soil (1)

$$K_a=\frac{1-\sin 38°}{1+\sin 38°}=0.24, K_p=\frac{1}{0.24}=4.17$$

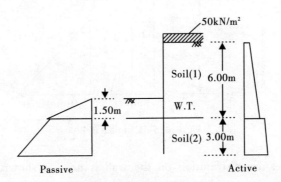

Fig. 8.4.6 Example 8.3

Table 8.4.1 Calculation procedure for Example 8.3

Soil	Depth (m)	Pressure (kN/m²)	
Active pressure			
1	0	0.24 ×50	=12.0
1	6	(0.24 ×50) + (0.24 ×18 ×6)	=37.9
2	6	0.36(50 + (18 ×6)) − (2 ×10 × $\sqrt{0.36}$)	=44.9
2	9	0.36(50 + (18 ×6)) − (2 ×10 × $\sqrt{0.36}$) + (0.36 ×10.2 ×3)	=55.9
Passive pressure			
1	0	0	
1	1.5	4.17 ×18 ×1.5	=112.6
2	1.5	(2.78 ×18 ×1.5) + (2 ×10 × $\sqrt{2.78}$)	=108.4
2	4.5	(2.78 ×18 ×1.5) + (2 ×10 × $\sqrt{2.78}$) + (2.78 ×10.2 ×3)	=193.5

The pressures in soil (1) are calculated using $K_a = 0.24$, $K_p = 4.17$ and $\gamma = 18$ kN/m³. Soil 1 is then considered as a surcharge of (18×6) kN/m² on soil (2), in addition to the surface surcharge. The pressures in soil (2) are calculated using $K_a = 0.36$, $K_p = 2.78$ and $\gamma' = (20 - 9.8) = 10.2$ kN/m³ (see Table 8.4.1). The active and passive pressure distributions are shown in Fig. 8.4.6. In addition, there is equal hydrostatic pressure on each side of the wall below the water table.

The Rankine theory will now be applied to cases in which the soil surface slopes at a constant angle β to the horizontal. It is assumed that the active and passive pressures act in a direction parallel to the sloping surface. Consider a rhombic clement of soil, with sides vertical and at angle β to the horizontal, at depth z in a semi-infinite mass. The vertical stress and the active or passive pressure are each inclined at β to the appropriate sides of the element, as shown in Fig. 8.4.7(a). Since these stresses are not normal to their respective planes (i.e. there are shear components), they are not principal stresses.

In the active case, the vertical stress at depth z on a plane inclined at angle β to the horizontal is given by

$$\sigma_z = \gamma z \cos \beta$$

and is represented by the distance OA on the stress diagram (Fig. 8.4.7(b)).

If lateral expansion of the soil is sufficient to induce the state of plastic equilibrium, the Mohr circle representing the state of stress in the element must pass through point A (such that the greater part of the circle lies on the side of A towards the origin) and touch the failure envelope for the soil. The active pressure pa is then represented by OB (numerically equal to

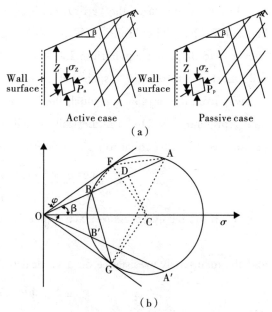

Fig. 8.4.7 Active and passive states for sloping surface

OB') on the diagram. When $c = 0$ the relationship between p_a and σ_z, giving the active pressure coefficient, can be derived from the diagram

$$K_a = \frac{p_a}{\sigma_z} = \frac{OB}{OA} = \frac{OB'}{OA} = \frac{OD - AD}{OD + AD}$$

Now

$$OD = OC \cos \beta$$

$$AD = \sqrt{(OC^2 \sin^2 \varphi - OC^2 \sin^2 \beta)}$$

Therefore

$$K_a = \frac{\cos \beta - \sqrt{(\cos^2 \beta - \cos^2 \varphi)}}{\cos \beta + \sqrt{(\cos^2 \beta + \cos^2 \varphi)}} \qquad (8.4.8)$$

Thus the active pressure, acting parallel to the slope, is given by

$$p_a = K_a \gamma z \cos \beta \qquad (8.4.9)$$

and the total active thrust on a vertical wall surface of height H is

$$E_a = \frac{1}{2} K_a \gamma H^2 \cos \beta \qquad (8.4.10)$$

In the passive case, the vertical stress σ_z is represented by the distance OB' in Fig. 8.4.7 (b). The Mohr circle representing the state of stress in the element, after a state of plastic equilibrium has been induced by lateral compression of the soil, must pass through B' (such that the greater part of the circle lies on the side of B' away from the origin) and touch the failure envelope. The passive pressure p_p is then represented by OA' (numerically equal to OA) and when $c = 0$ the passive pressure coefficient (equal to p_p / σ_z) is given by

$$K_p = \frac{\cos \beta + \sqrt{(\cos^2 \beta - \cos^2 \varphi)}}{\cos \beta - \sqrt{(\cos^2 \beta + \cos^2 \varphi)}} \qquad (8.4.11)$$

Then the passive pressure, acting parallel to the slope, is given by

$$p_p = K_p \gamma z \cos \beta \qquad (8.4.12)$$

and the total passive resistance on a vertical wall surface of height H is

$$E_p = \frac{1}{2} K_p \gamma H^2 \cos \beta \qquad (8.4.13)$$

The active and passive pressures can, of course, be obtained graphically from Fig. 8.4.7 (b). The above formulae apply only when the shear strength parameter

c is zero; when c is greater than zero the graphical procedure should be used.

The directions of the two sets of failure planes can be obtained from Fig. 8.4.7 (b). In the active case, the coordinates of point A represent the state of stress on a plane inclined at angle β to the horizontal, therefore point B' is the origin of planes, also known as the pole. (A line drawn from the origin of planes intersects the circumference of the circle at a point whose coordinates represent the state of stress on a plane parallel to that line.) The state of stress on a vertical plane is represented by the coordinates of point B. Then the failure planes, which are shown in Fig. 8.4.7 (a), are parallel to $B'F$ and $B'G$ (F and G lying on the failure envelope). In the passive case, the coordinates of point B' represent the state of stress on a plane inclined at angle β to the horizontal, and therefore point A is the origin of planes; the state of stress on a vertical plane is represented by the coordinates of point A'. Then the failure planes in the passive case are parallel to AF and AG.

Referring to Eqs. (8.4.8) and (8.4.11), it is clear that both K_a and K_p become equal to unity when $\beta = \varphi$; this is incompatible with real soil behaviour. Use of the theory is inappropriate, therefore, in such circumstances.

Example 8.4

A vertical wall 6m high, above the water table, retains a 20° soil slope, the retained soil having a unit weight of 18kN/m³; the appropriate shear strength parameters are $c' = 0$ and $\varphi' = 40°$. Determine the total active thrust on the wall and the directions of the two sets of failure planes relative to the horizontal.

olution

In this case the total active thrust can be obtained by calculation. Using Eq. (8.4.8),

$$K_a = \frac{\cos 20° + \sqrt{(\cos^2 20° - \cos^2 40°)}}{\cos 20° - \sqrt{(\cos^2 20° + \cos^2 40°)}} = 0.265$$

$$E_a = \frac{1}{2}K_a \gamma H^2 = \frac{1}{2} \times 0.265 \times 18 \times 6^2 \times 0.940 = 81 \text{ kN/m}$$

The result can also be determined using a stress diagram (Fig. 8.4.8). Draw the failure envelope on the τ/σ plot and a straight line through the origin at 20° to the horizontal. At a depth of 6 m,

$$\sigma_z = \gamma z \cos\beta = 18 \times 6 \times 0.940 = 102 \text{kN/m}$$

and this stress is set off to scale (distance OA) along the 20° line. The Mohr circle is then drawn as in Fig. 8.4.8 and the active pressure (distance OB or OB') is scaled from the diagram, i.e.

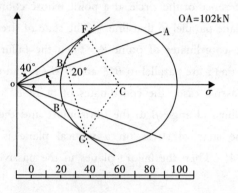

Fig. 8.4.8　Example 8.4

$$p_a = 27 \text{kN/m}^2$$

Then

$$E_a = \frac{1}{2} p_a H = \frac{1}{2} \times 27 \times 6 = 81 \text{kN/m}$$

The failure planes are parallel to $B'F$ and $B'G$ in Fig. 8.4.8. The directions of these lines are measured as 59° and 71°, respectively, to the horizontal (adding up to $90° + \varphi$).

8.5　Coulomb's theory of earth pressure

Coulomb's theory (1776) involves consideration of the stability, as a whole, of the wedge of soil between a retaining wall and a trial failure plane. The force between the wedge and the wall surface is determined by considering the equilibrium of forces acting on the wedge when it is on the point of sliding either up or down the failure plane, i. e. when the wedge is in a condition of limiting equilibrium. Friction between the wall and the adjacent soil is taken into account. The angle of friction between the soil and the wall material, denoted by δ, can be determined in the laboratory by means of a direct shear test. At any point on the wall surface a shearing resistance per unit area of $p_n \tan\delta$ will be developed, where p_n is the normal pressure on the wall at that point. A constant component of shearing resistance or 'wall adhesion', c_w, can also be assumed if appropriate in the case of clays. Due to wall friction the shape of the failure surface is curved near the bottom of the wall in both the active and passive cases, as indicated in Fig. 8.5.1, but in the Coulomb theory the failure surfaces are assumed to be plane in each case. In the active case, the curvature is slight and the error involved in assuming a plane surface, is relatively small. This is also true in the passive case for values of δ less than $\varphi/3$, but for the higher values of δ normally appropriate in practice the error becomes relatively large.

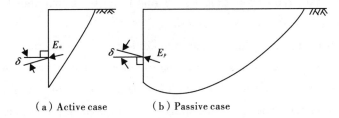

(a) Active case   (b) Passive case

Fig. 8.5.1 Curvature due to wall friction

The Coulomb theory is now interpreted as an upper bound plasticity solution; collapse of the soil mass above the chosen failure plane occurring as the wall moves away from or into the soil. Thus, in general, the theory underestimates the total active thrust and overestimates the total passive resistance (i. e. upper bounds to the true collapse loads). When $\delta = 0$, the Coulomb theory gives results which are identical to those of the Rankine theory for the case of a vertical wall and a horizontal soil surface, i. e. the solution for this case is exact because the upper and lower bound results coincide.

8.5.1 Active case

Figure. 8.5.2 (a) shows the forces acting on the soil wedge between a wall surface AB, inclined at angle α to the horizontal, and a trial failure plane BC, at angle θ to the horizontal. The soil surface AC is inclined at angle β to the horizontal. The shear strength parameter c will be taken as zero, as will be the case for most backfills. For the failure condition, the soil wedge is in equilibrium under its own weight (W), the reaction to the force (E) between the soil and the wall, and the reaction (R) on the failure plane. Because the soil wedge tends to move down the plane BC at failure, the reaction E acts at angle δ below the normal to the wall. (If the wall were to settle more than the backfill, the reaction E would act at angle δ above the normal.) At failure, when the shear strength of the soil has been fully mobilized, the direction of R is at angle φ below the normal to the failure plane (R being the resultant of the normal and shear forces on the failure plane). The directions of all three forces, and the magnitude of W, are known, and therefore the triangle of forces (Fig. 8.5.2 (b)) can be drawn and the magnitude of E determined for the trial in question.

A number of trial failure planes would have to be selected to obtain the maximum value of E, which would be the total active thrust on the wall. However, using the sine rule, E can be expressed in terms of W and the angles in the triangle of forces. Then the maximum value of E, corresponding to a particular value of θ, is given by $\partial E/\partial \theta = 0$. This leads to the following solution for E_a:

$$E_a = \frac{1}{2} K_a \gamma H^2 \qquad (8.5.1)$$

$$K_a = \left(\frac{\frac{\sin(\alpha - \varphi)}{\sin \alpha}}{\sqrt{\sin(\alpha + \delta)} + \sqrt{\frac{\sin(\varphi + \delta)\sin(\varphi - \beta)}{\sin(\alpha - \beta)}}} \right)^2 \qquad (8.5.2)$$

The point of application of the total active thrust is not given by the Coulomb theory but is assumed to act at a distance of $H/3$ above the base of the wall.

The Coulomb theory can be extended to cases in which the shear strength parameter c is greater than zero (i.e. c_u in the undrained case or if tangent parameter c' is used in the drained case). A value is then selected for the wall adhesion parameter c_w. It is assumed that tension cracks may extend to a depth z_0, the trial failure plane (at angle θ to the horizontal) extending from the heel of the wall to the bottom of the tension zone, as shown in Fig. 8.5.3. The forces acting on the soil wedge at failure are as follows:

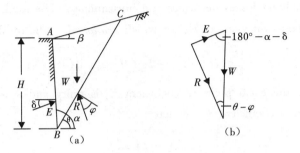

Fig. 8.5.2 Coulomb theory: active case with $c = 0$

(1) The weight of the wedge (W).

(2) The reaction (E) between the wall and the soil, acting at angle δ below the normal.

(3) The force due to the constant component of shearing resistance on the wall ($C_w = c_w \times EB$).

(4) The reaction (R) on the failure plane, acting at angle φ below the normal.

(5) The force on the failure plane due to the constant component of shear strength ($C = c \times BC$).

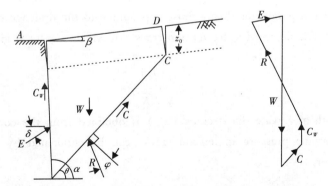

Fig. 8.5.3 Coulomb theory: active case with $c > 0$

The directions of all five forces are known together with the magnitudes of W, C_w and C, and therefore the value of E can be determined from the force diagram for the trial failure plane. Again, a number of trial failure planes would be selected to obtain the maximum value of E.

The special case of a vertical wall and a horizontal soil surface will now be considered. For the undrained condition ($\varphi_u = 0$), an expression for E can be obtained by resolving forces vertically and horizontally. The total active thrust is given by the maximum value of E, for which $\partial E/\partial \theta = 0$. The resulting value is

$$E_a = \frac{1}{2}\gamma(H^2 - z_0^2) - 2c_u(H - z_0)\sqrt{(1 + \frac{c_w}{c_u})}$$

For $\varphi_u = 0$ the earth pressure coefficient K_a is unity and it is convenient to introduce a second coefficient K_{ac}, where

$$K_{ac} = 2\sqrt{(1 + \frac{c_w}{c_u})}$$

For the fully drained condition in terms of tangent parameters c' and φ', it can be assumed that

$$K_{ac} = 2\sqrt{K_a(1 + \frac{c_w}{c'})}$$

In general, the active pressure at depth z can be expressed as

$$p_a = K_a \gamma z - K_{ac} c \qquad (8.5.3)$$

where

$$K_{ac} = 2\sqrt{K_a(1 + \frac{c_w}{c})} \qquad (8.5.4)$$

the shear strength parameters being those appropriate to the drainage conditions of the problem. The depth of a dry tension crack (at which $p_a = 0$) is given by

$$z_0 = \frac{2c\sqrt{(1 + \frac{c_w}{c})}}{\gamma \sqrt{K_a}} \qquad (8.5.5)$$

The depth of a water-filled crack (z_{0w}) is obtained from the condition $p_a = \gamma_w z_{0w}$. Hydrostatic pressure in tension cracks can be eliminated by means of a horizontal filter.

8.5.2 Passive case

In the passive case, the reaction E acts at angle δ above the normal to the wall surface (or δ below the normal if the wall were to settle more than the adjacent soil) and the reaction R at angle φ above the normal to the failure plane. In the triangle of forces, the angle between W and E is $180° - \alpha + \delta$ and the angle between W and R is $\theta + \varphi$. The total passive resistance, equal to the minimum value of E, is given by

$$E_p = \frac{1}{2} K_p \gamma H^2 \tag{8.5.6}$$

$$K_p = \left(\frac{\frac{\sin(\alpha + \varphi)}{\sin \alpha}}{\sqrt{\sin(\alpha - \delta)} + \sqrt{\frac{\sin(\varphi + \delta)\sin(\varphi + \beta)}{\sin(\alpha - \beta)}}} \right)^2 \tag{8.5.7}$$

However, in the passive case it is not generally realistic to neglect the curvature of the failure surface and use of Eq. (8.5.7) overestimates passive resistance, seriously so for the higher values of φ, representing an error on the unsafe side. It is recommended that passive pressure coefficients derived by Caquot and Kerisel (1966) should be used. Caquot and Kerisel derived both active and passive coefficients by integrating the differential equations of equilibrium, the failure surfaces being logarithmic spirals. Coefficients have also been obtained by Sokolovski (1965) by numerical integration.

In general, the passive pressure at depth z can be expressed as

$$p_p = K_p \gamma z - K_{pc} c \tag{8.5.8}$$

$$K_{pc} = 2 \sqrt{K_p \left(1 + \frac{c_w}{c}\right)} \tag{8.5.9}$$

8.6 Application of earth pressure theory to retaining walls

In the Rankine theory the state of stress in a semi-infinite soil mass is considered, the entire mass being subjected to lateral expansion or compression.

However, the movement of a retaining wall of finite dimensions cannot develop the active or passive state in the soil mass as a whole. The active state, for example, would be developed only within a wedge of soil between the wall and a failure plane passing through the lower end of the wall and at an angle of $45° + \varphi/2$ to the horizontal, as shown in Fig. 8.6.1 (a); the remainder of the soil mass would not reach a state of plastic equilibrium. A specific (minimum) value of lateral strain would be necessary for the development of the active state within the above wedge. A uniform strain within the wedge would be produced by a rotational movement ($A'B$) of the wall, away from the soil, about its lower end and a deformation of this type, of sufficient magnitude, constitutes the minimum deformation requirement for the development of the active state. Any deformation configuration enveloping $A'B$, for example, a uniform translational movement $A'B'$, would also result in the development of the active state. If the deformation of the wall were not to satisfy the minimum deformation requirement, the soil adjacent to the wall would not reach a state of plastic equilibrium and the lateral pressure would be between the active and at rest values. If the wall were to deform by rotation about its upper end (due, for example, to restraint by a prop), the conditions for the complete development of the active state would not be satisfied because of inadequate strain in the soil near the surface; consequently, the pressure near the top of the wall would approximate to the at rest value.

In the passive case the minimum deformation requirement is a rotational movement of the wall, about its lower end, into the soil. If this movement were of sufficient magnitude, the passive state would be developed within a wedge of soil between the wall and a failure plane at an angle of $45° + \varphi/2$ to the vertical as shown in Fig. 8.6.1 (b). In practice, however, only part of the potential passive resistance would normally be mobilized. The relatively large deformation necessary for the full development of passive resistance would be unacceptable, with the result that the pressure under working conditions would be between the at rest and passive values (and consequently providing a factor of safety against passive failure).

The selection of an appropriate value of φ' is of prime importance in the passive case. The difficulty is that strains vary significantly throughout the soil mass and in particular along the failure surface. The effect of strain, which is governed by the mode of wall deformation, is neglected both in the failure criterion and in analysis. In the earth pressure theories, a constant value of φ' is assumed throughout the soil above the failure surface whereas, in fact, the mobilized value of φ' varies. In the case of dense sands the average value of φ' along the failure surface, as the passive condition is approached, corresponds to a point beyond the peak on the stress-strain curve; use of the peak value of φ' would result, therefore, in an overestimation of passive resistance. It should be noted, however, that peak values of φ' obtained from triaxial tests are normally less than the corresponding values in plane strain, the latter being relevant in most retaining wall problems. In the case of loose sands, the wall deformation required to mobilize the ultimate value of φ' would be unacceptably large in practice.

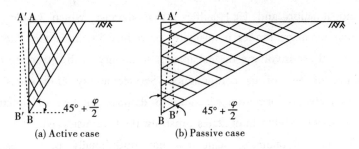

Fig.8.6.1 Minimum deformation conditions

The values of lateral strain required to mobilize active and passive pressures in a particular case depend on the value of K_0, representing the initial state of stress, and on the subsequent stress path, which depends on the construction technique and, in particular, on whether backfilling or excavation is involved in construction. In general, the required deformation in the backfilled case is greater than that in the excavated case for a particular soil. It should be noted that for backfilled walls, the lateral strain at a given point is interpreted as that occurring after backfill has been placed and compacted to the level of that point.

8.7 Design of earth-retaining structures

There are two broad categories of retaining structures: (1) gravity, or freestanding walls, in which stability is due mainly to the weight of the structure; (2) embedded walls, in which stability is due to the passive resistance of the soil over the embedded depth and, in most cases, external support. According to the principles of limit state design, an earth-retaining structure must not (a) collapse or suffer major damage, (b) be subject to unacceptable deformations in relation to its location and function and (c) suffer minor damage which would necessitate excessive maintenance, render it unsightly or reduce its anticipated life. Ultimate limit states are those involving the collapse or instability of the structure as a whole or the failure of one of its components. Service ability limit states are those involving excessive deformation, leading to damage or loss of function. Both ultimate and service ability limit states must always be considered.

The design of retaining structures has traditionally been based on the specification of a factor of safety in terms of moments, i.e. the ratio of the resisting (or restoring) moment to the disturbing (or overturning) moment. This is known as a lumped factor of safety and is given a value high enough to allow for all the uncertainties in the analytical method and in the values of soil parameters. It must be recognized that relatively large deformations are required for the mobilization of available passive resistance and that a structure could be deemed to have failed due to excessive deformation before reaching a condition of collapse. The approach, therefore, is to base design on ultimate limit states with the incorporation of an appropriate factor of safety to satisfy the requirements of service ability limit states. In general, the higher the factor of safety, the lower will be the deformation required to mobilize the proportion of passive resistance necessary for stability.

The stability of gravity (or freestanding) walls is due to the self-weight of the wall, perhaps aided by passive resistance developed in front of the toe. The traditional gravity wall (Fig. 8.7.1 (a)), constructed of masonry or mass concrete, is uneconomic because the material is used only for its dead weight. Reinforced concrete cantilever walls (Fig. 8.7.1 (b)) are more economic because the backfill itself, acting on the base, is employed to provide most of the required dead weight. Other types of gravity structure include gabion and crib walls (Fig. 8.7.1 (c) and (d)). Gabions are cages of steel mesh, rectangular in plan and elevation, filled with particles generally of cobble size, the units being used as the building blocks of a gravity structure. Cribs are open structures assembled from precast concrete or timber members and enclosing coarse-grained fill, the structure and fill acting as a composite unit to form a gravity wall.

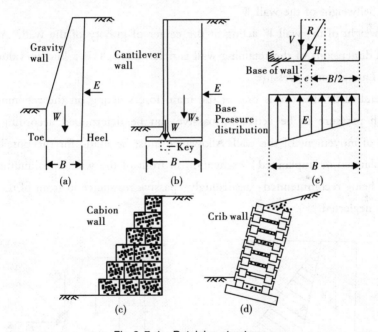

Fig. 8.7.1 Retaining structures

Limit states which must be considered in wall design are as follows:

(1) Overturning of the wall due to instability of the retained soil mass.

(2) Base pressure must not exceed the ultimate bearing capacity of the supporting soil, the maximum base pressure occurring at the toe of the wall because of the eccentricity and inclination of the resultant load.

(3) Sliding between the base of the wall and the underlying soil.

(4) The development of a deep slip surface which envelops the structure as a whole.

(5) Soil and wall deformations which cause adverse effects on the wall itself or on adjacent structures and services.

(6) Adverse seepage effects, internal erosion or leakage through the wall: consideration should be given to the consequences of the failure of drainage systems to operate as intended.

(7) Structural failure of any element of the wall or combined soil/structure failure.

The first step in design is to determine all the forces on the wall.

(1) Selfweight of the wall W

Selfweight of the wall W acting at the center of gravity of the wall. After the type and dimensions of the retaining wall is determined, W is a known value.

(2) Lateral earth pressure

Lateral earth pressure is one of the main forces acting on the retaining wall. The earth pressure to be active or passive can be determined according to the direction of movement of the wall. Allowance must be made for the possibility of future (planned or unplanned) excavation in front of the wall, a minimum depth of 0.5m being recommended: accordingly, passive resistance in front of the wall is normally neglected.

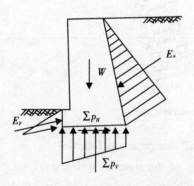   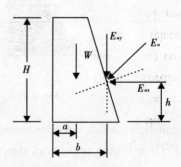

ig.8.7.2 Forces acting on the wall   Fig.8.7.3 Stability assessment of the wall

(3) Reaction force acting on the base of the wall

Reaction force acting on the base of the wall can be divided into horizontal and vertical components. The distribution of the vertical components is assumed to be the same as a eccentrically loaded foundation, i.e. distributed as a trapezoid. The resultant vertical force is represented using $\sum p_V$ which acting at the centroid of the trapezoid. The resultant horizontal force is represented by $\sum p_H$ (Fig. 8.7.2).

8.7.1 Anti-sliding stability verification

Dividing the earth pressure E_a into two components

$$E_{ax} = E_a \cos(\varepsilon + \delta) \quad (8.7.1)$$
$$E_{ay} = E_a \sin(\varepsilon + \delta) \quad (8.7.2)$$

then the anti-sliding stability can be calculated using

$$K_s = \frac{(W + E_{ay})\mu}{E_{ax}} \geqslant 1.3 \quad (8.7.3)$$

where, K_s is factor of safety of the anti-sliding stability, a minimum of 1.3 of which is specified by the *Code for design of building foundation* (GB 50007 −2011). μ is coefficient of friction of the base of the wall, which can be determined by tests.

8.7.2 Anti-overturning stability verification

The anti-overturning stability of the wall can be verified by ensuring that the total resisting moment about the toe exceeds the total overturning moment (Fig. 8.7.3).

$$K_t = \frac{Wa + E_{ay}b}{E_{ax}h} \geqslant 1.5 \quad (8.7.4)$$

where K_t is the factor of safety of anti-overturning stability, a minimum of 1.5 of which is specified by the *Code for design of building foundation* (GB 50007 −2011).

8.7.3 Bearing capacity verification

Bearing capacity of a retaining wall can be verified using the same method as used in eccentrically loaded foundations. The stress at the base of the wall should fulfil the following equations.

$$\frac{1}{2}(\sigma_{max} + \sigma_{min}) \leqslant f \qquad (8.7.5)$$

$$\sigma_{max} \leqslant 1.2f \qquad (8.7.6)$$

where σ_{max} and σ_{min} are maximum and minimum stress acting on the base of the wall, f is the bearing capacity of the soil underlying the base of the wall.

Example 8.5

Details of retaining wall are shown in Fig. 8.7.4, the water table being below the base of the wall. The height of the wall is $H = 6.0$ m and the back of the wall inclined at an angle of $\alpha = 100°$. The unit weight of the backfill is 18.5kN/m^3. The surface of the backfill inclined at an angle of $\beta = 10°$. Characteristic values of the shear strength parameters for the backfill are $c = 0$ and $\varphi = 30°$. The angle of friction between the wall and the backfill is $\delta = 20°$. The coefficient of friction between bottom of the wall and the foundation is $\mu = 0.4$. Bearing capacity of the supporting soil is $f = 180$ kPa. Please design the wall.

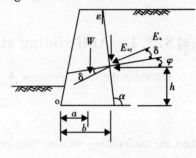

Fig.8.7.4　Example 8.5

Solution

(1) Estimate the dimension of the wall

Let the width of the top and bottom surface to be 1.0m and 5.0m, respectively, thus the self weight of the wall will be

$$W = \frac{(1.0 + 5.0)H\gamma_c}{2} = 432 \text{kN/m}$$

(2) Calculate the earth pressure

Because $\alpha = 100°$, $\beta = 10°$, $\varphi = 30°$, $\delta = 20°$, from Equation 8.5.2, we can obtain $K_a = 0.46$.

According to Equation 8.5.1

$$E_a = \frac{1}{2}\gamma H^2 K_a = 153 \text{kN/m}$$

The vertical component of the earth pressure is

$$E_{ay} = E_a \sin(\varepsilon + \delta) = 76.5 \text{kN/m}$$

The horizontal component of the earth pressure is

$$E_{ax} = E_a \cos(\varepsilon + \delta) = 132.3 \text{kN/m}$$

(3) Anti-sliding stability verification

According to Equation 8.7.3

$$K_s = \frac{(W + E_{ay})\mu}{E_{ax}} = 1.54 > 1.3 \text{ (safe)}$$

The factor of safety is too large. Let the width of the bottom surface to be 4.0 m, and then self weight of the wall $W' = \frac{(1.0\text{m} + 4.0\text{m})H\gamma}{2} = 360\text{kN/m}$

$$K_s = \frac{(W' + E_{ay})\mu}{E_{ax}} = 1.32 > 1.3 \text{ (appropriate)}$$

(4) Anti-overturning stability verification

The lever arms for the forces against the toe of the wall O are

Lever arm for W': $a = 2.17$ m;

Lever arm for E_{ay}: $b = 3.65$ m;

Lever arm for E_{ax}: $h = 2.00$ m.

Using Eq. (8.7.4)

$$K_t = \frac{Wa + E_{ay}b}{E_{ax}h} = 4.0 \geqslant 1.5 \text{ (safe)}$$

(5) Bearing capacity verification

① The force acting on the foundation is

$$N = W' + E_{ay} = 436.5 \text{kN/m}$$

② Distance from the acting point of the resultant force to the toe O

$$x = \frac{Wa + E_{ay}b - E_{ax}h}{N} = 1.82\text{m}$$

③ Eccentricity

$$e = \frac{B}{2} - x = 0.18$$

Therefore

$$\begin{Bmatrix} P_{max} \\ P_{min} \end{Bmatrix} = \frac{N}{K_t}(1 \pm \frac{6e}{B}) = \begin{Bmatrix} 138.6 \\ 79.6 \end{Bmatrix} \text{kPa}$$

$$\frac{1}{2}(P_{max} + P_{min}) = 109.1 \text{kPa} < f = 180 \text{kPa}$$

$$P_{max} = 138.6 \text{kPa} < 1.2f = 216 \text{kPa}$$

The bearing capacity of the supporting soil is satisfactory.

EXERCISES

8.1 The backfill behind a retaining wall, located above the water table, consists of a sand of unit weight $17 kN/m^3$. The height of the wall is 6m and the surface of the backfill is horizontal. Determine the total active thrust on the wall according to the Rankine theory if $c' = 0$ and $\varphi' = 37°$. If the wall is prevented from yielding, what is the approximate value of the thrust on the wall?

8.2 Plot the distribution of active pressure on the wall surface shown in Fig. 8.1. Calculate the total thrust on the wall (active + hydrostatic) and determine its point of application. Assume $\delta = 0$ and $c_w = 0$.

8.3 The section through a gravity retaining wall is shown in Fig. 8.2, the unit weight of the wall material being $23.5 kN/m^3$. The unit weight of the backfill is $19 kN/m^3$ and design values of the shear strength parameters are $c' = 0$ and $\varphi' = 36°$. The value of δ between wall and backfill and between base and foundation soil is $25°$. Use the limit state method to determine if the design of the wall is satisfactory with respect to the overturning and sliding limit states.

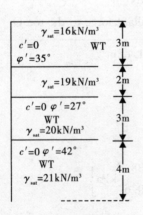

Fig. 8.1   Exercise 8.2

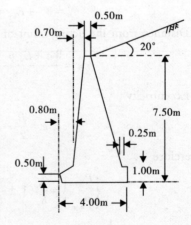

Fig. 8.2   Exercise 8.3

REFERENCES

1. Caquot, A. and Kerisel, J. (1966). Traite de Mecanique des Sols (the fourth edition) [M]. Gauthier-Villars, Paris.
2. Ingold, T. S. (1979). The Effects of Compaction on Retaining Walls [J]. Geotechnique, 29(3), 265 −283.
3. Jaky, J. (1944). The Coefficient of Earth Pressure at Rest [J]. Journal of the Society of Hungarian Architects and Engineers, 7, 355 −358.
4. Massarsch, K. R. (1979). Lateral Earth Pressure in Normally Consolidated clay [C]. Proceedings of the Seventh European Conference on Soil Mechanics and Foundation Engineering, Brighton, England, 2, 245 −250.
5. Rankine, W. M. J. (1857). On Stability on Loose Earth [C]. Philosophic Transactions of Royal Society, London, Part I, 9 −27.
6. Sherif, M. A., Fang, Y. S. and Sherif, R. I. (1984). K_a and K_0 Behind Rotating and Non-Yielding Walls [J]. Journal of Geotechnical Engineering, ASCE, 110 (GT1), 41 −56.
7. Sokolovski, V. V. (1965). Statics of Granular Media [M]. Pergamon Press, Oxford.
8. Terzaghi, K., Peck, R. B. and Mesri, G. (1996). Soil Mechanics in Engineering Practice (the third edition) [M]. John Wiley & Sons, New York.

APPENDIX I

English-Chinese translation of frequently-used words

A

active earth pressure	主动土压力
adamic earth	红黏土
additional stress	附加应力
allowable bearing capacity	容许承载力
allowable settlement	容许沉降
angle of internal friction	内摩擦角
angle of repose	休止角
artesian water	承压水
artificial fills	人工填土
axial load	轴向荷载

B

backfill	回填土
base of slope	坡底
bearing capacity factor	承载力系数
bearing capacity of ground	地基承载力
bound water	结合水
bulk density	密度
bulk modulus	体积模量
buoyant unit weight	浮重度

C

capillarity	毛细作用
capillary pressure	毛细管压力

APPENDIX I

capillary water	毛细水
centric load	中心荷载
circular slip surface	圆弧滑动面
clay	黏粒/黏土
coarse aggregate	粗骨料
coarse sand	粗砂
coefficient of active earth pressure	主动土压力系数
coefficient of consolidation	固结系数
coefficient of curvature	曲率系数
coefficient of earth pressure at rest	静止土压力系数
coefficient of lateral pressure	侧压力系数
coefficient of passive earth pressure	被动土压力系数
coefficient of permeability	渗透系数
coefficient of vertical consolidation	竖向固结系数
cohesion	黏聚力
cohesive soil	黏性土
collapsibility	湿陷性
collapsible loess	湿陷性黄土
compaction curve	击实曲线
compaction test	击实试验
compactness	密实度
compressibility	压缩性
compression curve	压缩曲线
compression index	压缩指数
concentrated load	集中荷载
cone penetrometer for liquid limit test	锥式液限仪
confined compression test	侧限压缩试验
consistency index	稠度指数
consistency	稠度
consolidated quick direct shear test	固结快剪试验
consolidated undrained triaxial test, CU	固结不排水三轴试验
consolidated-drained triaxial test, CD	固结排水三轴试验

consolidation settlement	固结沉降
constant head permeability test	常水头渗透试验
constrained diameter	限制粒径
controlled-strain triaxial test	应变控制式三轴试验
corner-points method	角点法
Coulomb's theory of earth pressure	库仑土压力理论
critical edge pressure	临塑荷载
critical hydraulic gradient	临界水力梯度
crushed stone	碎石

Darcy's law	达西定律
degree of consolidation	固结度
degree of saturation	饱和度
depth of foundation	基础埋深
deviator stress	偏应力
dilatancy	剪胀性
direct shear apparatus	直剪仪
direct shear test	直剪试验
discharge velocity	渗透速度
disturbed samples	扰动土样
drainage	排水
dry density	干密度
dry unit weight	干重度
Duncan-Chang model	邓肯-张模型

earth pressure at rest	静止土压力
earth pressure	土压力
eccentric load	偏心荷载
eccentricity	偏心距
effective angle of internal friction	有效内摩擦角

effective cohesion	有效黏聚力
effective diameter	有效粒径
effective normal stress	有效法向应力
effective principal stress	有效主应力
effective stress path	有效应力路径
effective stress	有效应力
effective unit weight	有效重度
envelope of Mohr's circles	摩尔圆包线
excess pore water pressure	超静孔隙水压力
expansion, swelling index	回弹指数
expansive soil	膨胀土

factor of compaction	压实系数
factor of safety	安全系数
failure envelope	破坏包线
falling-head permeability test	变水头渗透试验
field loading test	现场载荷试验
fine sand	细砂
flexible foundation	柔性基础
flocculent structure	絮凝结构
flow line	流线
flow net	流网
flow path	流径
free water	自由水
frozen soil	冻土

general-shear failure	整体剪切破坏
geostatic stress	自重应力
gradation	级配
grading curve	级配曲线

grain diameter/particle size	粒径
grain size distribution curve	粒径级配曲线
grain size distribution	粒径级配
grain size fraction/grain group	粒组
grainage	粒度
gravel	砾石
gravitational water	重力水
gravity retaining wall	重力式挡土墙
groundwater table	地下水位

honeycomb structure	蜂窝结构
hydraulic conductivity	透水性
hydraulic gradient	水力梯度
hydrostatic pressure	静水压力

immediate settlement	瞬时沉降
in situ testing	原位试验
initial hydraulic gradient	起始水头梯度
instability	失稳
intact specimen	原状土样
internal scour	潜蚀

laminar flow	层流
landslide	滑坡
lateral geostatic stress	侧向自重应力
layered soil	成层土
layerwise summation method	分层总和法
line load	线荷载
liquefied sand	流砂

liquid limit apparatus	液限仪
liquid limit	液限
liquidity index	液性指数
loading plate test	荷载试验
load-settlement curve	荷载－沉降曲线
local shear failure	局部剪切破坏
loessial soil	黄土类土
loess	黄土

maximum dry density	最大干密度
mean diameter	平均粒径
method of slices	条分法
modulus of compressibility	压缩模量
modulus of deformation	变形模量
Mohr's circle	摩尔圆
Mohr-Coulomb criterion	摩尔－库仑准则
mucky soil	淤泥质土

natural ground	天然地基
natural slope	天然边坡
natural void ratio	天然孔隙比
natural water content	天然含水量
non-cohesive soil	无粘性土
non-uniform settlement	不均匀沉降
normal stress	法向应力
normally consolidated soil	正常固结土

oedometer test	固结试验
oedometric modulus	侧限压缩模量

one-dimensional consolidation	单向固结
optimum water content	最优含水量
organic soil	有机质土
overconsolidated soil	超固结土
overconsolidation ratio	超固结比

particle composition	颗粒组成
particle density	土粒密度
passive earth pressure	被动土压力
peak strength	峰值强度
peaty soil	泥炭质土
peat	泥炭
pedogenesis	成土作用
permeability	渗透性
plain fill	素填土
plastic limit	塑限
plasticity index	塑性指数
Poisson's ratio	泊松比
pore pressure coefficient	孔隙压力系数
pore water pressure	孔隙水压力
porosity	孔隙率
preconsolidation pressure	前期固结应力
principle of effective stress	有效应力原理
proportional limit load	比例极限荷载
punching shear failure	冲切破坏

Q

quantity of flow	流量
quick shear test	快剪试验

Rankine's theory of earth pressure	朗肯土压力理论

rate of shear	剪切速率
rebound curve	回弹曲线
relative surface area	比表面积
remolded sample	重塑土样
residual deformation	残余变形
retaining structure	挡土结构
retaining wall	挡土墙

S

saline soil	盐渍土
sand	砂粒/砂
saturated density	饱和密度
saturated soil	饱和土
saturated unit weight	饱和重度
secondary consolidation settlement	次固结沉降
seepage flow	渗流
seepage force	渗流力
settlement	沉降
shear failure	剪切破坏
shear modulus	剪切模量
shear plane, shear surface	剪切面
shear strain	切应变
shear strengthparameter	抗剪强度参数
shear strength	抗剪强度
shear stress	切应力
shear test	剪切试验
shrinkage limit	缩限
silt fraction	粉粒粒组
silty clay	粉质粘土
silty sand	粉砂
silty soil	粉质土
silt	粉粒/粉土

single grain fabric	单粒结构
slip surface	滑裂面
slope angle	坡角
slope stability	边坡稳定性
slope toe	坡趾
slope	边坡
slow shear test	慢剪试验
sludge	淤泥
soft clay	软黏土
soft foundation	软弱地基
soil mechanics	土力学
stability coefficient	稳定系数
state of limit equilibrium	极限平衡状态
strain hardening	应变硬化
strain path	应变路径
strain softening	应变软化
stress dispersion	应力扩散
stress history	应力历史
stress path	应力路径
stress-strain relationship	应力-应变关系
stress	应力
strip foundation	条形基础
strip load	条形荷载
submerged sample	浸水试样
submersion	浸水
surface water	地表水
Swedish circle method	瑞典圆弧法

T

tangent modulus	切线模量
Terzaghi's theory of one-dimensional consolidation	太沙基一维固结理论
total head	总水头

APPENDIX I

total normal stress	总法向应力
total principal stress	总主应力
total stress approach	总应力法
total stress path	总应力路径
triaxial apparatus	三轴仪
triaxial compression test	三轴固结试验
triaxial shear test	三轴剪切试验
truly triaxial test apparatus	真三轴试验仪
turbulent flow	紊流

ultimate bearing capacity	极限承载力
ultimate load	极限荷载
ultimate settlement	最终沉降量
unconfined compression strength	无侧限抗压强度
unconfined compression test	无侧限抗压试验
unconsolidated undrained triaxial test, UU	不固结不排水三轴试验
under-consolidated soil	欠固结土
uniform settlement	均匀沉降
unit weight	重度
unsaturated soil	非饱和土

vane shear apparatus	十字板剪力仪
vane shear test	十字板剪切试验
vertical stress increase coefficient	竖向附加应力系数
virgin compression curve	原始压缩曲线
void ratio	孔隙比
volumetric strain	体积应变

water content/moisture content	含水量

• • • 227

Young's modulus 弹性模量

APPENDIX II

Answers to the exercises

Chapter 1

1.5 $0.972, 1.37 \text{g/cm}^3$

1.6 $1.8 \text{g/cm}^3, 1.61 \text{g/cm}^3, 12\%, 40\%, 47.6\%$

1.7   169.6kg

1.8   Medium dense

1.9 $15.7 \text{kN/m}^3, 19.7 \text{kN/m}^3, 9.9 \text{kN/m}^3, 18.7 \text{kN/m}^3, 19.3\%$

Chapter 2

2.5 $7.5 \times 10^{-2} \text{cm/s}$

2.6   1.02

2.7   The heaving sand happens.

2.8   5.76m

Chapter 3

3.1   0, 35kPa, 53kPa, 72kPa, 93kPa

3.2 $\sigma_{zB} = 8.24 \text{kPa}$

3.3 $\sigma_{z=1\text{m}} = 252.49 \text{kPa}, \sigma_{z=2\text{m}} = 140.37 \text{kPa}, \sigma_{z=3\text{m}} = 82.10 \text{kPa},$
      $\sigma_{z=4\text{m}} = 53.58 \text{kPa}, \sigma_{z=6\text{m}} = 29.17 \text{kPa}, \sigma_{z=8\text{m}} = 19.12 \text{kPa}$

3.4 $\sigma_{zA} = 0.0806p, \sigma_{zO} = 0.4136p$

3.5 $p_{\max} = 331.67 \text{kPa}, p_{\min} = 65.0 \text{kPa}$

3.6   81.3kPa

Chapter 4

4.1 $\alpha_{1-2}=0.09\text{MPa}^{-1}$, $E_{S1-2}=21.6\text{MPa}$, soil is low compressible.
4.2 23.1cm
4.3 194.4days
4.5 156.6mm, 166mm
4.6 $6.33\times10^{-3}\text{cm}^2/\text{s}$

Chapter 5

5.1 Failure
5.2 $c=0, \varphi=41.8°$

Chapter 6

6.1 (1) $p_{cr}=82.6\text{kPa}$, $p_{\frac{1}{4}}=89.7\text{kPa}$, $p_{\frac{1}{3}}=92.1\text{kPa}$; (2) $p_{cr}=82.6\text{kPa}$, $p_{\frac{1}{4}}=85.8\text{kPa}$, $p_{\frac{1}{3}}=86.9\text{kPa}$
6.2 (1) 226.3kPa, 90.5kPa; (2) 259kPa, 291.8kPa

Chapter 7

7.1 1.64, 1.47
7.2 1.13
7.3 A slip will occur.
7.4 1.08

Chapter 8

8.1 76.5kN/m, 122kN/m
8.2 571kN/m, point of application is 8.75m from the top of the wall.
8.3 The bearing capacity of the supporting soil is not satisfactory.

后 记

本英文教材按照我国土力学课程的结构体系和最新规范编写而成。进入新世纪以来，教育部倡导在具备条件的高校可采取双语教学方式。为了适应新世纪以来高等学校本科土建类专业不断改革的教学计划和课程体系，提高综合素质的人才培养模式的要求，作者根据长期从事土力学教学的经验，结合英文版教材的编写方式，在符合本门学科要求的同时，重构了英文《土力学》课程体系和教学内容。编写时注重所传授知识的系统性和逻辑性，内容条理性清晰，采用国内的最新规范标准，便于读者的学习和理解。

本英文教材与高等教育出版社新世纪《土力学》（第2版）（赵树德、廖红建主编）、新体系《土力学》（廖红建、柳厚祥主编）的中文教材基本对应，在教学和学习中可同时使用或参考。

本书由西安交通大学廖红建教授组织编写，成员有西安建筑科技大学苏立君教授、新疆大学肖正华副教授、西安交通大学李杭州讲师，他们当中大多数有留学经历，并取得博士学位和从事博士后研究。各章编写情况如下：绪论、第一章、第二章、附录：廖红建；第三章和第四章：肖正华、廖红建、李杭州；第五章：廖红建、李杭州；第六章：李杭州；第七章和第八章：苏立君；全书由廖红建教授、李杭州讲师整理统稿。西安交通大学岩土工程研究生郝东瑞协助做了不少工作，在此表示衷心感谢。

本书由西安交通大学俞茂宏教授审阅，并在内容组织等方面提出了许多宝贵的建议，在此谨致以衷心的感谢。

本书得到了西安交通大学"十二五"规划教材项目以及国家自然科学基金项目（41172276，51279155）的大力支持，在此表示衷心的感谢。

限于编者的水平，书中难免有不妥之处，恳请广大读者批评指正。

编 者

2015年7月